EASY AI: CONQUER PROMPT ENGINEERING

PERFECTING THE ART OF TALKING TO MACHINES

MICHAEL GRANT MALLOY

© **Copyright 2025 - All rights reserved.**

The content contained within this book may not be reproduced, duplicated or transmitted without direct written permission from the author or the publisher.

Under no circumstances will any blame or legal responsibility be held against the publisher, or author, for any damages, reparation, or monetary loss due to the information contained within this book, either directly or indirectly.

Legal Notice:

This book is copyright protected. It is only for personal use. You cannot amend, distribute, sell, use, quote or paraphrase any part, or the content within this book, without the consent of the author or publisher.

Disclaimer Notice:

Please note the information contained within this document is for educational and entertainment purposes only. All effort has been executed to present accurate, up to date, reliable, complete information. No warranties of any kind are declared or implied. Readers acknowledge that the author is not engaged in the rendering of legal, financial, medical or professional advice. The content within this book has been derived from various sources. Please consult a licensed professional before attempting any techniques outlined in this book.

By reading this document, the reader agrees that under no circumstances is the author responsible for any losses, direct or indirect, that are incurred as a result of the use of the information contained within this document, including, but not limited to, errors, omissions, or inaccuracies.

CONTENTS

Preface	7
1. INTRODUCTION: Why Prompt Engineering Matters Now	9
2. Understanding Prompts: The Building Blocks of AI Communication	17
3. Prompt Structures and Frameworks That Work	26
4. Context and Constraints – Setting Boundaries for Better Results	36
5. Iteration and Refinement – Mastering the Feedback Loop	43
6. Advanced Prompting Techniques	51
7. Prompt Engineering Across Domains	59
8. Common Mistakes and How to Avoid Them	67
9. Ethics and Responsibility in Prompt Design	75
10. The Future of Prompt Engineering – From Art to Automation	83
11. Prompt Engineering Success Stories	89
Study & Discussion Guide	102
Prompt Library	108
References	119

AUTHORS NOTE

When I first began writing these books to try and make AI less frightening to the average professional without a lot of IT experience, I quickly realized that this book would be necessary. And then I realized it may be the most important one. Because learning to use AI effectively, no matter what your role with AI might be, comes down to how well you are able to communicate your wants, needs, or desires, to the AI tool you happen to be using. There is both a science and an art to doing this successfully.

Thank you for joining me on this journey. May the pages ahead equip you with tools, inspire new ideas, and empower you to write your own success stories with AI.

--Michael Grant Malloy

PREFACE

We are living in a moment of profound change. In just a few short years, artificial intelligence has moved from the realm of research labs and science fiction into our classrooms, businesses, hospitals, and even living rooms. AI writes, summarizes, analyzes, and creates at a scale and speed we have never experienced before.

But here's the truth: the value you will get from AI depends more on the clarity of your questions than on the complexity of the AI system itself.

That is why prompt engineering has emerged as one of the most important new skills of the 21st century. Prompt engineering is about learning to communicate effectively with intelligent systems: to guide them, to structure their responses, to anticipate their limitations, and to harness their strengths.

Across these chapters, you'll learn frameworks for crafting effective prompts, see real-world examples from multiple industries, and reflect on the ethical responsibilities that come with shaping AI outputs. I've also included guiding questions throughout, designed to help you pause, consider, and apply what you learn directly to your own context.

My aim is to help you engage with AI more skillfully regardless of your direction or profession.

CHAPTER 1
INTRODUCTION: WHY PROMPT ENGINEERING MATTERS NOW

IN LESS THAN A YEAR, more than 100 million people around the world have started using AI tools like ChatGPT—a number that would have sounded unimaginable just a few short seasons ago. This sudden wave isn't just happening in tech circles; everywhere you look—offices, startups, classrooms, and coffee shops—digital professionals are experimenting with prompts, searching for an edge, or simply trying to keep pace as the world shifts faster than ever before.

If you've ever stared at a blinking cursor, unsure how to get what you really want from an AI tool, or felt buried under a mountain of buzzwords and conflicting advice, you're not alone. The truth is, even the most digitally fluent feel overwhelmed by the sheer speed of change. Maybe you've tried prompting only to get generic or off-target results, wondered if this technology was meant for coders alone, or worried that you're falling behind. I see you. This book was written to turn that frustration into clarity, and confidence, so you can seize these new opportunities without hesitation and stress.

Imagine crafting prompts that don't just 'work,' they transform how you do business, spark clever ideas, streamline your daily tasks, and even open doors to brand new income streams. Picture the hours won back from tedious busywork, the pride of sharing polished reports and proposals, or the thrill of surprising your team (or your boss) with AI-powered insights and creative energy. You'll learn to design precise prompts that unlock hidden value from today's smartest AI systems, boost productivity, enhance research efficiency, and embed ethical practices in every task you automate—with frameworks and resources at your fingertips throughout.

My own journey began in the trenches: learning AI solutions by the seat of my pants, translating technology breakthroughs into plain English, and guiding teams through rapid technology pivots I barely had time to teach myself. After a decade bridging the gap between brilliant engineers and everyday users, I've learned that anyone willing to ask sharp questions, and quickly refine and reiterate these questions, can master prompt engineering—no PhD required. My goal is to be your translator and collaborator, making advanced strategies approachable, actionable, and always relevant to your real-world goals.

Here's the secret that most people overlook. AI doesn't work in isolation. It works in response to you. The words you use to communicate with AI, i.e. your prompts, shape the quality, relevance, and accuracy of the results you get. In fact, your success with AI tools often depends less on the sophistication of the technology and more on the precision of your questions and/or instructions.

At its core, prompt engineering is the art and science of asking better questions. If you know how to send a smart email, brief a colleague, or outline a project, you're halfway there already. With straightforward templates, live examples, and a little curiosity, anyone can learn to guide AI effectively, freeing you from guesswork and replacing confusion with control. Prompt engineering is a skill you can practice, refine, and master, and it's one that will only grow in value as AI becomes increasingly integrated into the way we work and live.

To make your journey rewarding, this book is built for hands-on progress and busy lives. Whether you lead teams, consult clients, launch side hustles, or tackle creative projects, you'll find targeted tools tailored to where you stand today, with plenty of support as you grow.

What sets this book apart is its focus on tomorrow's skills, not just text-based prompts but next-generation methods like Chain of Thought, Retrieval-Augmented Generation, and multimodal strategies that blend words, images, and even code. These approaches help you future-proof your workflow, stay competitive, and build mastery as AI capabilities evolve. As you explore, you'll gain the edge needed to outpace change rather than play catch-up.

By the time you finish this book, you'll be able to:

- Craft prompts that reliably produce accurate and high-quality results.
- Adapt your prompting strategies to different industries and tasks.
- Recognize when to add context, when to set boundaries, and when to let creativity flow.
- Avoid the pitfalls that lead to wasted time and frustrating outputs.
- Think like a "prompt engineer" — not just a tool user, but a partner who knows how to unlock AI's full potential.

THE RAPID RISE OF GENERATIVE AI

As highlighted in this chapter's opening, the scale of AI adoption is unlike anything business has seen before. Just a year ago, few could have predicted that over 75% of knowledge workers would be tapping into generative AI tools at work—and nearly half of them picked up these skills in only the last six months. The speed and reach of this adoption wave creates both anxiety and excitement. For professionals reading this, AI isn't something happening quietly in the background. It's showing up in daily workflows, client demands, and even conversations with peers about career futures. This moment—this sudden, collective leap—forms the backdrop for understanding why prompt engineering now matters as much as technical or communication skills ever have.

Look around any office or virtual workspace: You'll spot someone running an AI-powered chatbot for support tickets, a marketer drafting copy in minutes using large language models, or a sales manager prepping targeted outreach packs without touching a spreadsheet. The shift feels like the landline-to-smartphone moment—but compressed into just a handful of quarters. Businesses across every sector are watching project cycles shrink as content creation, document analysis, data wrangling, and proposal writing get automated overnight. In technology teams, developers harness AI to supercharge coding; elsewhere, HR uses it to scan resumes quickly, while supply chain managers optimize routes and inventory with algorithms that learn on the fly.

This new landscape directly affects jobs and expertise. Imagine a manager who used to spend hours collating reports; now, custom prompts deliver clear executive summaries in minutes. A consultant automates routine research, then pivots to shaping client strategy. Freed from repetitive tasks, people re-focus on creative problem solving, relationship building, and big-picture planning. But the transition carries unease: If AI handles more administrative work, where does human value come from? What happens when job postings start to list "prompt writing" as a must-have skill? The reality is that roles are changing fast.

Workplace expectations are catching up to this change. Projects once expected by week's end now arrive overnight. Teams regularly review AI-generated deliverables alongside traditional work, and having a strong handle on AI-powered tools is becoming standard. Some employers emphasize AI literacy in their hiring; others reward employees who can shape effective prompts that drive real results. It's common to see skills like "experience with generative AI tools" or "prompt engineering ability" highlighted in job descriptions, giving early adopters a measurable edge.

Competence in prompt engineering stands out as the new secret weapon for gaining an advantage. Professionals who invest time learning how to create effective prompts are already landing better projects, serving clients faster, and customizing AI outputs in ways generic users can't match. Mastering this art levels the playing field, letting newcomers leapfrog veterans who stick with old approaches.

The key drivers behind this shift include:

- Ease of use- the emergence of no-code AI platforms allows everyone to participate
- Affordability- new platforms have sliding scale pricing structures for small and mid-size enterprises
- Open access platforms allow micro-enterprises or individuals to take advantage

QUALITY OUTPUT

Thanks to user-friendly interfaces, access to the services provided by AI platforms is now relatively simple. Almost anyone can start a project or automate a task with a few clicks. Still, the ease of entry into AI tools often leads to a false sense of confidence. Just because most people can figure out how to write simple prompts doesn't mean everyone will be able to prompt well without effort. And most people assume that using AI successfully is about having the most advanced system. But here's the surprising truth: *two people can use the same AI tool and get completely different results.* The difference isn't in the software; it's in the prompt. And without skillful prompts opportunities may be missed. This is why prompt engineering isn't a gimmick. It's a skill that should be given the respect it's due.

Here's a simplistic example:

- Prompt A: "Write about climate change."
- Prompt B: "Write a 300-word article explaining the impact of climate change on coastal cities, using simple language suitable for high school students."

Prompt A will give you a general, unfocused response. Prompt B, however, provides clarity, context, constraints, and audience guidance — and the result will almost always be sharper, more relevant, and more useful.

One of the simplest ways to see the power of prompts is through side-by-side comparisons.

Bad Prompt: "Help me with marketing."

Likely Output: A generic list of strategies: "Use social media, email campaigns, and paid ads."

Better Prompt: "Act as a marketing consultant. Suggest five low-cost digital strategies for a small business selling handmade jewelry online, focusing on Instagram and TikTok."

Likely Output: Tailored strategies like "Post behind-the-scenes videos of jewelry creation on TikTok" or "Run Instagram Story polls to engage customers."

The difference is night and day. The second prompt doesn't just ask for help — it sets a role, a context, and specific context. That's the essence of prompt engineering.

CREATIVITY LOST

Aside from the potential to slow down workflow, weak prompting skills can also stifle creativity, both individually or within a team. The spark of creativity often fizzles when AI returns the obvious rather than the original. Yet, with expert prompting, the same tools can help generate unexpected connections, angles, or breakthroughs in thoughts, concepts, or designs. The possibilities multiply exponentially when AI is inspired with clear, thoughtful, and well-designed prompts.

A simple shift in wording can transform a dull response into something imaginative and unique. Consider how this example demonstrates that prompts might be not only instructions but also creative sparks:

- "Write a poem about dogs." (Generic output.)
- "Write a humorous haiku from the perspective of a dog waiting for its owner to return home." (Creative, surprising, and fun output.)

Skillful prompts allow you to mold AI into an assistant, a collaborator, or even a creative partner.

Competitive Advantage

Teams that master skillful prompting launch products sooner, adjust strategy faster, and communicate clearer—staking claims in their respective market ahead of those organizations that are still untangling their AI confusion. Teams that haven't mastered skillful prompting waste time and resources endlessly tweaking and revising AI prompts, redoing reports, missing milestones, and becoming disillusioned and distrustful of their AI tools. The cost of lagging behind isn't limited to day-to-day headaches—it creates openings for competition to surge ahead in digital transformation and AI-driven workflows. Without prompt engineering mastery, falling behind is not a distant worry. That risk is already reshaping professional landscapes everywhere.

The Science and the Art

Prompt engineering sits at the intersection of science and art. The "science" comes from understanding how AI interprets instructions and applying structured methods — clarity, specificity, context. The "art" comes from experimentation, iteration, and knowing when to loosen the boundaries to let creativity flourish.

For example, a data analyst might use prompts like:

- "Summarize this dataset in three bullet points highlighting unusual trends."

Meanwhile, a novelist might use prompts like:

- "Generate three alternate endings for a suspense novel where the detective is secretly the villain."

Both approaches rely on the same principles, but their application looks different depending on the domain.

REAL-WORLD APPLICATIONS

The importance of prompt engineering extends far beyond writing. Consider these industries:

- **Business:** Leaders use prompts to generate reports, analyze competition, and draft communication.
- **Healthcare:** Clinicians experiment with prompts for summarizing patient notes or researching treatment guidelines.
- **Education:** Teachers create custom lesson plans or quizzes by carefully shaping prompts.
- **Creative Arts:** Writers, artists, and designers use prompts to spark new ideas or refine drafts.

In each case, the difference between a vague instruction and a well-structured prompt can mean hours of saved time, reduced frustration, and better results.

A Skill Anyone Can Learn

Perhaps the most encouraging truth is you don't need a technology background to master prompt engineering. Unlike coding, which requires years of technical training, prompt engineering is accessible to anyone who knows how to communicate clearly. If you can ask a question, you can learn to craft a skillful prompt. This is why businesses are already looking for "prompt engineers" -because the role does not require any type of degree. All it takes to be an exceptional Prompt Engineer is a blend of curiosity, clarity, and creativity.

The Road Ahead

Think of this book as your map and toolkit. Each chapter adds another tool, another insight, another practice you can use to unlock the full potential of AI. By the end, you won't just be using AI; you'll be guiding it with precision and creativity, turning every interaction into a powerful collaboration.

CHAPTER 2
UNDERSTANDING PROMPTS: THE BUILDING BLOCKS OF AI COMMUNICATION

HAVE you ever thought that talking to AI was as simple as typing in quick questions or commands without giving it much thought? Have you been treating your AI tools with the same approach you've used for google searches over the last decade? Many people do, assuming the process is straightforward and casual. Yet, this assumption usually leads to lackluster answers from AI tools—responses that miss the mark in both business meetings and creative brainstorms. When prompts are vague or rushed, the result is missed chances for deeper insights, faster workflows, or surprising innovations. Understanding the true nature of prompts is like finding a secret code; it's what transforms basic interaction into powerful communication.

This chapter takes on the common myth head-on by delving into what prompts actually are and why they hold such importance. It guides you through seeing prompts not just as typed words, but as carefully crafted bridges connecting your goals with AI's capabilities. By mastering this foundational skill, you'll unlock new ways to harness AI effectively across work, teaching, and creativity. We'll start by breaking down the essence of prompts themselves, setting you up to communicate clearly and confidently with AI every time.

WHAT PROMPTS ACTUALLY ARE

Before we tackle how to structure a prompt, we need to define what a prompt really is and illustrate how this concept underpins all meaningful AI communication. A prompt is the message, question, or instruction you type into an AI, telling it what to do. Every AI interaction starts with a prompt. This can look like a simple question— "What time is it in Tokyo?"—a direct command— "Summarize this article"— or a richer set of instructions giving context, style preferences, or rules — "Write an email update to my team about our project status based on the last meeting minutes. Write the email using a friendly but concise tone." These varied prompts highlight one core idea: each AI session begins with your **intent** translated into words. That single phrase, set of directions, or thoughtful description becomes the bridge between your needs and the output you receive.

Imagine someone at work types "meeting" into an AI tool. The result might be vague or unrelated to their goal. Now picture them entering, "Summarize key tasks and deadlines from today's product meeting minutes—highlight who's responsible for each action." Suddenly the output becomes useful. Being clear and detailed means less back-and-forth editing, quicker insights, and outputs tailored to actual business demands. That difference saves time when writing reports, handling customer emails, or preparing lesson plans. When you provide clarity and details—what, who, format, tone—the AI produces relevant responses that make life easier, help meet deadlines, and free up energy for creative thinking instead of reworking lackluster results.

You don't need to learn code or special syntax to communicate with AI. Unlike programming, where you build everything from the ground up with strict rules and technical language, prompting uses everyday words. Writing a prompt is like briefing a colleague—you tell them what you want, they interpret and deliver. With programming, a set of instructions always gives the same result unless you rewrite the code; with prompting, you adapt your questions as you see the AI's replies, adjusting your input in real-time until the answer fits your needs. This flexibility lets you experiment and refine without any fear of breaking the system. Even if you're not tech-savvy, you already possess the crucial skill for effective prompting: the ability to communicate in plain language. Anyone who can give clear instructions or ask precise questions is ready to succeed as a prompt engineer.

CORE PRINCIPLES OF EFFECTIVE PROMPTING

Prompt engineering rests on four foundational pillars: **clarity, specificity, context, and structure**. Think of these as the grammar of working with AI. Master them, and you can reliably produce sharper, more accurate, and more relevant responses across almost any domain.

I. Clarity: The First Rule of Effective Prompting

If AI is like a mirror of human communication, then clarity is the difference between seeing your reflection clearly or through a fogged-up glass. Unclear prompts confuse the system, leading to vague, off-track, or generic answers.

Example:

- Unclear: "Write something about leadership."
- Clear: "Write a 500-word article explaining three qualities that make great leaders, using real-world examples from sports and business."

The first leaves AI guessing what you want. The second eliminates ambiguity by spelling out *what, how long, how many,* and *in what style*.

Tips for Clarity

- Avoid vague verbs like *do, write about,* or *help with*. Replace them with precise instructions like *summarize, analyze, compare,* or *generate*.
- Break large requests into smaller, manageable steps.
- Use plain language — don't assume the AI understands jargon unless you define it.

📌 **Guiding Question:** *If I gave this prompt to a person with no background knowledge, would they know exactly what to deliver?*

II. Specificity: Narrowing the Scope for Accuracy

AI models are capable of generating information across countless domains, but their breadth is also a weakness. Without guardrails, they may wander into irrelevance. Specificity acts like a funnel, narrowing down the range of possible answers so the AI focuses on what matters most.

Example:

- Broad: "Tell me about World War II."
- Specific: "Summarize the role of women in the U.S. workforce during World War II in 300 words, focusing on cultural and economic changes."

Notice how the second version not only narrows the topic but also specifies length, perspective, and angle. This keeps the AI from dumping a history textbook's worth of information and instead delivers a focused, useful summary.

Dimensions of Specificity

- **Length:** "In two paragraphs" or "List five strategies."
- **Format:** "As a table," "as a bulleted list," or "as a persuasive essay."
- **Perspective:** "Explain as if you are a teacher talking to 5th graders" or "from the viewpoint of a startup founder."
- **Domain:** "Within healthcare" or "specific to digital marketing."

📌 **Guiding Question:** *What boundaries can I set to keep the AI focused on exactly what I need?*

III. Context: The Secret Ingredient

Clarity and specificity are powerful, but without context, you're still leaving too much to chance. Context is the background information that frames the AI's task — the "why" and "who" behind the request.

Think of context as stage-setting. If you were hiring a freelancer, you wouldn't just say, "Write me an article." You'd explain who the audience is, what the goals are, and how the piece will be used. AI requires the same.

Example:

- Low-context: "Explain blockchain."
- High-context: "You are writing for an audience of small-business owners with no technical background. Explain blockchain in simple terms, focusing on how it could help them reduce transaction costs."

The first example will likely give a technical or overly complex answer. The second frames the audience and purpose, which allows the AI to adjust tone, vocabulary, and focus.

How to Add Context

- Identify the **audience**: Who will read or use this output?
- Define the **purpose**: Is this to inform, persuade, entertain, or analyze?
- Provide the **situation**: Where will this be used? A blog post? A presentation? A class?
- Add **constraints**: Such as reading level, tone (professional, casual, humorous), or format.

📌 **Guiding Question:** *If I don't give this context, will the AI make assumptions that could lead to irrelevant results?*

IV. Structure: Building a Framework for the AI

The final principle is structure — the way you organize your instructions. Even with clarity, specificity, and context, an unstructured prompt can leave the AI uncertain about priorities. Structure helps the AI "see" the order and hierarchy of tasks.

Why Structure Matters

AI doesn't always know which parts of your request are most important. For example:

- "Write a detailed blog post about the benefits of meditation and include scientific research, beginner tips, and a motivational conclusion in under 1,000 words."

That's a lot to process in one sentence. The AI may overemphasize research while neglecting beginner tips, or vice versa.

Now compare:

1. Write a blog post under 1,000 words.
2. Include three sections:
    - Scientific research on the benefits of meditation.
    - Practical beginner tips.
    - A motivational conclusion that inspires readers to start meditating today.

This structured approach ensures the AI hits every requirement in order.

Structuring Techniques

- **Numbered lists** for multi-step instructions.
- **Bulleted lists** to specify required elements.
- **Step-by-step prompting** (ask for one part at a time, then build).
- **Scaffolding**: Start with a broad outline, then fill in details in later prompts.

📌 **Guiding Question:** *Does this prompt present information in a way that makes it easy for AI to prioritize and execute?*

PUTTING THE PRINCIPLES TOGETHER

These four principles aren't meant to work in isolation. The real power comes when you combine them.

Weak Prompt: "Write about healthy eating."

Strong Prompt (with all four principles):

"You are a nutrition coach writing for busy parents. Create a 700-word article that explains five simple, budget-friendly strategies for healthy eating. Structure the article with an introduction, body paragraphs for each strategy, and a conclusion with a motivational call-to-action."

Let's break it down:

i. **Clarity:** Clear task — write an article.
ii. **Specificity:** 700 words, five strategies.
iii. **Context:** Audience is busy parents, written by a "nutrition coach."
iv. **Structure:** Introduction, body, conclusion, call-to-action.

The result? A polished, audience-appropriate piece instead of a generic nutrition essay.

Case Study: The CEO's Dilemma

To illustrate how these principles work in practice, let's look at a real-world scenario.

A CEO wants to use AI to draft a memo about adopting remote work policies. Here's the initial attempt:

Prompt 1: "Write a memo about remote work."

Output: A bland, general overview of remote work benefits and drawbacks.

After learning about prompt engineering, the CEO refines it:

Prompt 2:

"You are the CEO of a mid-sized technology company. Write a one-page internal memo to staff announcing a new hybrid work policy. The memo should:

1. Explain the reasoning behind the decision (employee well-being, productivity, and flexibility).
2. Provide specific details (3 days in-office, 2 days remote).
3. Emphasize company culture and teamwork.
4. End with a motivational note about adapting to change."

Output: A professional, persuasive memo that hits every point clearly.

By applying clarity, specificity, context, and structure, the CEO turned an average AI draft into a polished internal communication.

Why These Principles Work

At this point, you might be wondering: *Why do these principles make such a difference?* The answer lies in how AI models are trained. They predict text based on patterns in massive datasets. Without constraints, the possibilities are infinite, which often leads to generic results. By applying these principles, you're essentially narrowing the prediction field so the AI selects the "right" path more consistently. In other words, you're reducing randomness and increasing relevance.

GUIDED PRACTICE

To make these principles stick, try this exercise:

1. Pick a topic (e.g., "customer service").
2. Write the vaguest prompt you can think of.
3. Apply clarity by specifying the task.
4. Add specificity by defining the scope, the format, and the perspective.
5. Layer in context by framing audience and purpose.
6. Apply structure by organizing the output requirements.

Compare the before-and-after outputs. You'll see immediately how much sharper the second version is.

The Human Element

Here's an important truth: mastering these principles doesn't just make AI better — it makes *you* better. These skills overlap with strong communication practices in everyday life. Being clear, specific, contextual, and structured are the same qualities that define good teaching, good management, and good leadership. Prompt engineering, then, isn't about technology. It's about learning to communicate with precision in a world where words increasingly shape outcomes.

Conclusion: The Foundation of Prompt Mastery

Clarity, specificity, context, and structure are the bedrock of effective prompting. They are simple in theory, but applying them consistently requires practice and intention. Once you internalize these principles, you'll stop thinking of AI as a mysterious black box and start treating it as a responsive collaborator.

In the chapters ahead, we'll build on these foundations with more advanced techniques: frameworks, iteration, personas, and multi-step reasoning. But no matter how complex things get, these four principles will remain your compass.

Every time you sit down to craft a prompt, ask yourself:

- Is it clear?
- Is it specific?
- Does it include context?
- Is it well-structured?

If the answer is yes, you're already halfway to unlocking the full power of AI.

CHAPTER 3
PROMPT STRUCTURES AND FRAMEWORKS THAT WORK

SO NOW YOU know the core pillars that are the foundation of effective prompting: clarity, specificity, context, and structure. But knowing the pillars isn't enough. One needs repeatable frameworks that ensure these pillars are used reliably when structuring a prompt. Frameworks act like maps or blueprints: they reduce guesswork, streamline your thinking, and ensure that your prompts consistently yield high-quality results.

At the same time, an open and adventurous attitude to creating your prompts allows for exploration, testing, and refining of prompts to ensure your AI tools learn your unique communication style. This is where concept of iteration comes into play, making prompt engineering as much an art as it is a science.

Since AI platforms are language learning models, they can and will learn from repetitive (iterative) use. Iteration is also important because rarely does the first version of any prompt yield a perfect result. Instead of accepting a mediocre response, or starting over from scratch, you can learn how making small, focused adjustments to your prompts will lead AI closer to the output you expect and desire.

Soon, in just a few short cycles, you'll watch vague outputs transform into crisp, compelling messaging, high quality images, or reliable predictions. For professionals working at speed, iteration offers rapid feedback without long delays, aligning perfectly with the reality of tight deadlines and evolving business needs. This dual approach of combining reliable frameworks with iterative refinement is the key that will unlock smoother collaboration and impressive outcomes from your AI tools.

Let's get started exploring the most effective prompt structures and frameworks, determine when and how to use them, as well as when and how to explore, tweak, or test them, and illustrate with examples from different domains. By the end, you'll have a toolkit of ready-to-use patterns for everything from quick queries to complex multi-step projects.

THE SCIENCE: BASIC PROMPT FRAMEWORKS

1. The Role–Task–Context–Output Framework

One of the most powerful and versatile frameworks is **Role + Task + Context + Output Format**. This structure tells the AI *who it should be, what it should do, under what conditions, and in what form the result should come.*

Template:

- **Role:** "Act as a [expert, teacher, consultant, etc.]."
- **Task:** "Do [specific action]."
- **Context:** "Given [background, audience, purpose]."
- **Output Format:** "Deliver results as [list, essay, table, etc.]."

Example:

"Act as a nutrition coach. Create a one-week meal plan for someone who wants to lose weight, given a daily calorie limit of 1,800 and vegetarian preferences. Deliver the results as a table with days, meals, and calorie counts."

This structure works because it leaves little room for misinterpretation.

Without it, the AI might give generic advice like "Eat more vegetables." With it, you get a concrete, useful plan.

2. Step-by-Step Prompting

Complex requests often overwhelm AI if given all at once. Step-by-step prompting breaks down the task into logical stages, reducing the chance of missing details.

Example (business report):

> **Step 1:** "Generate a list of five key trends in digital marketing for 2025."
> **Step 2:** "For each trend, summarize why it matters to small businesses."
> **Step 3:** "Turn this into a 1,000-word report with an introduction, body, and conclusion."

Each step builds on the previous one, ensuring accuracy and coherence. This is especially useful for long-form writing, coding, and multi-part projects.

3. Chain-of-Thought Prompting

Sometimes you want the AI not just to give you an answer, but to **show its reasoning process**. Chain-of-thought prompting explicitly asks for step-by-step logic.

Example (math problem):

- Prompt: "Explain step by step how to solve 48 ÷ (4 × 2). Show your reasoning before giving the final answer."
- Output: The AI explains order of operations, shows the calculation, then gives the answer (6).

This approach is powerful for problem-solving, decision-making, and teaching scenarios. It slows down the AI's "thinking," leading to more accurate and transparent results.

4. Few-Shot and Zero-Shot Prompting

Frameworks also differ in how much example data you provide.

- **Zero-shot prompting**: You give no examples — just instructions.
    - Prompt: "Summarize this paragraph in plain English."
- **Few-shot prompting**: You give one or more examples to "teach" the AI.
    - Prompt: "Here's an example summary: [short example]. Now summarize this new paragraph in the same style."

Few-shot prompting is especially useful when you need a **consistent style or format**, while zero-shot is faster for simpler tasks.

5. Persona-Based Prompting

Another effective structure is to assign the AI a **persona**. This helps it "roleplay" and deliver responses aligned with that perspective.

Examples:

- "Act as a lawyer reviewing a contract. Highlight potential risks."
- "Pretend you are a children's author. Write a bedtime story about kindness."
- "You are a coding instructor. Explain recursion to a beginner using simple metaphors."

By stepping into a role, the AI narrows its frame of reference, reducing the risk of generic or irrelevant answers.

6. Scaffolding Prompts

Scaffolding is the process of building progressively from simple to complex prompts. You start with an outline or framework, then refine with more detail.

Example (article writing):

1. "Give me an outline for a blog post on sustainable fashion."
2. "Now expand the outline into section summaries."
3. "Write the full 1,200-word article with sources."

This approach mirrors how humans work — first brainstorm, then structure, then draft. It helps manage long outputs and improves coherence.

7. Grounded Context Prompts

A Grounded Context Prompt is a prompt framework used in when explicitly instructing an AI system to base its response only on retrieved context (documents, passages, or data) provided in the prompt, also known as Retrieval-Augmented Generation (RAG).

The goal is to:

- Minimize hallucination (AI making up facts).
- Increase accuracy by tethering answers to external sources.
- Provide transparency when citing or referencing documents.

Example

Context (data) provided:

1. GLP-1 receptor agonists help lower blood sugar by enhancing insulin secretion.
2. Recent studies (2024) show GLP-1 drugs also support weight loss in type 2 diabetes patients.

Prompt: You are an AI assistant. Use only the information provided in the CONTEXT above to answer.

Question: What are the benefits of GLP-1 receptor agonists?

Model's Response (Grounded): GLP-1 receptor agonists lower blood sugar by increasing insulin secretion and have been shown to support weight loss in type 2 diabetes patients (Context 1 & 2).

Why this works:

- The AI stays within the given context.
- It avoids adding extra claims (e.g., about cardiovascular effects) unless those are in the context.
- It provides a clear, factual answer grounded in retrieved evidence.

PROMPT FRAMEWORKS FOR SPECIFIC GOALS

1. Explaining Concepts

- Framework: *Role + Audience + Task + Examples*
- Example: "Act as a high school science teacher. Explain the concept of photosynthesis to 10th-grade students using a simple analogy and two real-world examples."

2. Creative Writing

- Framework: *Genre + Style + Perspective + Constraints*
- Example: "Write a short story in the style of Edgar Allan Poe, from the perspective of a detective who uncovers his own crime. Limit it to 500 words."

3. Decision-Making

- Framework: *Context + Options + Criteria + Recommendation*
- Example: "Given a small business with $10,000 to spend on marketing, compare the pros and cons of Facebook ads, Google Ads, and influencer partnerships. Recommend the best option based on cost-effectiveness and audience reach."

4. Problem-Solving

- Framework: *Problem Statement + Step-by-Step Reasoning + Proposed Solution*

- Example: "A team's productivity has dropped by 20%. Explain possible causes step by step, then propose three solutions."

WHEN TO USE WHICH FRAMEWORK

Not all frameworks fit every situation. Here are some rules of thumb:

- Use **Role–Task–Context–Output** for most business, education, or structured writing tasks.
- Use **Step-by-Step or Scaffolding** when working on long projects or detailed analysis.
- Use **Chain-of-Thought** for reasoning-heavy problems.
- Use **Few-Shot** when style consistency is crucial.
- Use **Persona-Based** when creativity, tone, or perspective matters.
- Use **Grounded Context Prompts** when there is a need or reason to seek knowledge from a specific source.

The key is to treat frameworks as **tools in a toolbox**. Choose the one that best fits your purpose.

COMMON MISTAKES IN FRAMEWORK USE

- **Overloading:** Trying to cram too many frameworks into one prompt (e.g., persona + few-shot + step-by-step all at once) can confuse the AI.
- **Inconsistency:** Switching formats mid-prompt without clear transitions.
- **Lack of Testing:** Assuming the first attempt is final instead of refining.

Remember: frameworks are meant to simplify, not complicate.

CASE STUDY: THE TEACHER'S LESSON PLAN

A teacher wants to create a lesson plan on the water cycle. Here's the unstructured attempt:

- Prompt: "Write a lesson plan about the water cycle."
- Output: A basic explanation with little organization.

Now, using a framework:

- Prompt: "Act as a 5th-grade science teacher. Create a 45-minute lesson plan on the water cycle. Include:
  1. Learning objectives.
  2. Key vocabulary words.
  3. A short interactive activity.
  4. A five-question quiz."

The output is structured, age-appropriate, and classroom-ready — all because the framework guided the AI step by step.

Practice Exercise

Choose a task (e.g., "plan a vacation," "write a sales pitch," "analyze survey data"). Try three versions:

1. A simple prompt.
2. A Role–Task–Context–Output prompt.
3. A Step-by-Step or Persona-based prompt.

Compare the results. Which was most useful? Which felt easiest to refine?

This practice will help you instinctively choose the right framework for the job.

FRAMEWORKS AS PROMPTING SHORTCUTS

Frameworks aren't rigid rules — they're shortcuts to better thinking. They allow you to apply the principles of clarity, specificity, context, and structure in predictable ways. More importantly, they train you to think like a **prompt engineer**, reducing wasted effort and increasing the reliability of your results.

As AI grows more sophisticated, frameworks will remain your anchor. They ensure that no matter how complex your request, you can break it down into manageable parts and guide the AI to deliver exactly what you need.

THE ART: ITERATIVE IMPROVEMENT

Now that you have the basic frameworks for creating effective prompts, it's time to think about how you go about perfecting your craft, one prompt at a time.

- Begin by creating a relatively simple prompt in the "Role–Task–Context–Output" framework for a topic you already know well. Review what comes back.
- Now consider about what type of adjustments to the prompt might improve the AI's output.
- Revise the prompt to include the adjustments.
- Review what comes back
- Repeat this process until your output is how you envisioned it would or should be.

Now repeat this process for each type of framework, using topics you already know well so you can judge how your iterations are impacting the AI's output. This is a learning process for both you and AI, who is learning your language style. Consider this type of practice is similar to learning to perform any art form, from drawing to ballet to playing piano. Effective users expect to revise and clarify to get the best results from AI. By practicing, they learn to do so efficiently and effectively.

📌 **Guiding Question:** *Have I ever become proficient at anything without concerted effort and practice? If not, why should using AI be any different?*

Saving 'Prompt Libraries': Future-Proofing Your Efforts

Once you do land on solid prompts that might be valuable for repeated use, consider capturing them in a prompt library. Organize by project or use case—maybe a simple spreadsheet or digital notebook. Annotate each entry with notes about when and why it worked: "Great for quick executive recaps" or "Best used for technical blog ideas."

Add columns for rating effectiveness and jot down tweaks you'd like to try next time. Think of your prompt library as your kitchen spice rack—ready whenever you need flavor. This habit saves time and lets you build on your own progress, turning experiments into reliable resources.

CHAPTER 4
CONTEXT AND CONSTRAINTS – SETTING BOUNDARIES FOR BETTER RESULTS

IN THE PREVIOUS CHAPTER, you learned about prompt frameworks — the blueprints that guide AI to produce more structured and predictable outputs. Now, we dive deeper into two essential elements that make those frameworks work even better: **context** and **constraints**.

Think of prompt engineering as a conversation with a highly intelligent but easily distracted assistant. Without clear context, they don't know the "why" behind your request. Without constraints, they ramble on, miss the point, or overwhelm you with irrelevant detail. But when you set both context and constraints thoughtfully, you channel the AI's power in exactly the direction you need.

WHY CONTEXT MATTERS

AI systems are trained on massive amounts of data, but they don't "know" your situation unless you tell them. Context bridges that gap. It ensures that the AI understands not only the task but also the environment, purpose, and audience.

Consider this:

- Prompt without context: "Explain inflation."
- Prompt with context: "You are writing for college freshmen in an introductory economics class. Explain inflation in simple, everyday terms, and provide two examples from the past five years that illustrate its impact on consumer goods."

The second version transforms the answer. The AI now knows the *audience* (college freshmen), the *purpose* (teaching), the *tone* (simple terms), and even the *angle* (recent examples).

Types of Context

1. **Audience Context**: Who is this for? (Children, professionals, beginners, experts.)
2. **Purpose Context**: Why is this being written? (To inform, persuade, entertain, analyze.)
3. **Situational Context**: Where will it be used? (Presentation, email, academic paper, marketing campaign.)
4. **Cultural or Domain Context**: What background knowledge should the AI assume? (Healthcare, law, business, education.)

THE ART OF SETTING CONSTRAINTS

While context sets the stage, constraints set the boundaries. They tell the AI *what not to do* or *how far to go*. Constraints help you avoid overly broad or unusable results.

Examples of constraints include:

- **Length**: "In under 300 words."
- **Format**: "Provide as a bulleted list."
- **Tone**: "Keep it lighthearted and humorous."
- **Scope**: "Focus only on U.S. data since 2020."

Prompt without constraints: "Write about climate change."

Prompt with constraints: "Write a 400-word op-ed in a persuasive tone, arguing that local governments should prioritize renewable energy projects over fossil fuel subsidies. Include two data points from the U.S. after 2020."

The first prompt could lead to a vague essay, while the second is precise, actionable, and tailored.

FINDING THE BALANCE

Too much context or too many constraints can overwhelm the AI. For example:

"Act as a sustainability consultant. Write a 750-word report for mid-level executives in a U.K. construction company on reducing carbon emissions, with three case studies, five bulleted action steps, a glossary of terms, and an appendix formatted for Excel spreadsheets."

This prompt is so overloaded that the AI may produce something messy or incomplete. The trick is balance: give enough context and constraints to guide, but not so much that you drown the system in instructions.

When you have complex needs to communicate to AI that make prompt engineering challenging, it's important to remember that various frameworks are better suited for different scenarios. Take the time to explore which framework might work best for each unique situation. Experimenting with different frameworks can often be helpful in finding the most effective solution.

CONTEXT IN ACTION: BUSINESS EXAMPLE

A small business owner wants an AI-generated email campaign.

- **Weak Prompt:** "Write marketing emails."
- **Improved Prompt:** "You are a marketing consultant creating a three-email sequence for a small online bookstore. The

audience is young adults who enjoy fantasy novels. Keep each email under 200 words, use a playful tone, and end with a call-to-action to visit the online store."

The improved prompt uses context (audience, business type, tone) and constraints (email count, word limit, CTA) to ensure the output is useful right away.

CONSTRAINTS IN ACTION: HEALTHCARE EXAMPLE

A medical researcher needs a summary of recent studies.

- **Weak Prompt:** "Summarize diabetes research."
- **Improved Prompt:** "Summarize three peer-reviewed studies on Type 2 diabetes published since 2021. Limit your research to articles published in either JAMA or the Journal of Diabetes Research. Limit your explanation to 500 words. Focus only on clinical outcomes, not policy debates."

Here, the constraints (study count, year range, length, topic scope) keep the AI from pulling in irrelevant or outdated data.

TECHNIQUES FOR USING CONTEXT EFFECTIVELY

1. **Define the Role and Audience Together**
    - Example: "Act as a career coach speaking to recent college graduates entering the tech industry."
2. **Use Storytelling Context**
    - Example: "Pretend you are a historian in 2050 explaining how artificial intelligence changed global trade."
3. **Add Background Information**
    - Example: "Here's the company mission: [insert text]. Based on this, write a vision statement for the next five years."

TECHNIQUES FOR APPLYING CONSTRAINTS EFFECTIVELY

1. **Set Hard Limits**
   - "Explain in 150 words or less."
   - "Give me exactly five bullet points."
2. **Specify Format**
   - "Present as a markdown table with columns for challenge, solution, and outcome."
3. **Direct Tone and Style**
   - "Write in the voice of a motivational speaker."
   - "Keep the language suitable for 6th graders."
4. **Narrow Time or Scope**
   - "Use data from the past decade."
   - "Focus only on U.S. policies."

COMMON MISTAKES WITH CONTEXT AND CONSTRAINTS

- **Assuming the AI knows your audience:** Without specifying, the AI may default to general explanations.
- **Overloading with constraints:** Too many requirements can confuse the model.
- **Contradictory instructions:** Telling the AI to be both "formal" and "casual" in the same prompt.
- **Leaving out priorities:** If you list 10 requirements without ranking them, the AI may emphasize the wrong ones.

CASE STUDY: THE CONSULTANT'S REPORT

A management consultant uses AI to prepare a client report.

- **Prompt without context or constraints:**
- "Write a report about workforce productivity."
- **Result:** A generic overview with little relevance.
- **Prompt with context and constraints:**
- "Act as a management consultant preparing a briefing for the CEO of a manufacturing company. In 800 words, analyze three

key factors affecting workforce productivity since the shift to hybrid work models. Include one case study from the automotive sector and present recommendations in a bulleted format."
- **Result:** A sharp, relevant, client-ready briefing that fits the CEO's needs.

EXERCISE: REFINING PROMPTS

Try this exercise to practice adding context and constraints:

1. Start with a simple prompt: "Explain artificial intelligence."
2. Add context: "Explain artificial intelligence to high school students."
3. Add constraints: "Keep it under 250 words, use two real-world examples, and avoid technical jargon."

Compare the outputs. You'll see how each layer improves clarity and usefulness.

THE HUMAN PARALLEL

Interestingly, these same principles apply in everyday human communication. Imagine asking an intern to "make a presentation" without offering any additional instructions. Without providing context, the intern may struggle to understand the expectations. Without clear context, they won't know the intended audience or purpose of the presentation and might select an inappropriate topic. At the same time, having no constraints may result in the intern creating an overly elaborate 40-slide deck that takes an hour to present, even though you were expecting a concise one-page summary to be delivered in just five minutes. Communicating both elements clearly to the intern at the time of the request often results in a presentation that closely aligns with your expectations.

Prompt engineering is simply translating good communication practices into your interactions with AL.

Conclusion: Context + Constraints = Precision

Context and constraints are like the twin rails of a train track. Context tells the AI where it's going; constraints keep it from veering off the chosen track. Without them, prompts are vague, outputs are generic, and frustration follows. With them, prompts become powerful levers that channel AI's capabilities into targeted, relevant, and actionable results.

As you continue building your prompting skills, always ask yourself:

- Have I given the AI enough context to understand the task?
- Have I set the right constraints to keep the output focused?

In the next chapter, we'll explore how to **iterate and refine** prompts — the feedback loop that turns good results into great ones.

CHAPTER 5
ITERATION AND REFINEMENT – MASTERING THE FEEDBACK LOOP

BY NOW, you've learned how to apply clarity, specificity, context, structure, and frameworks to craft strong prompts. But here's the reality: **the first prompt you write is rarely the best one.** Just like writing, designing, or problem-solving in any field, the secret to great outcomes lies in iteration.

Iteration is the process of **refining prompts through feedback loops** — testing, adjusting, and improving until the AI's responses align with your needs. In other words, it's about treating AI not as a vending machine that dispenses answers, but as a partner in conversation.

This chapter explores why iteration matters, how to refine prompts effectively, and what strategies you can use to create a systematic feedback loop.

WHY ITERATION MATTERS

AI is probabilistic, not deterministic. That means its outputs aren't fixed — the same prompt may produce slightly different answers each time. Even when it's consistent, your initial request may not fully capture what you need. Iteration allows you to:

- **Clarify Misunderstandings:** If the AI misinterprets your request, refinement helps steer it back on track.
- **Add Missing Pieces:** The first draft might lack examples, structure, or depth. Iteration fills in the gaps.
- **Explore Alternatives:** Iteration lets you compare different approaches and choose the best.
- **Save Time Long-Term:** Instead of endlessly re-prompting from scratch, refinement helps you converge on usable results more efficiently.

Think of iteration as the editing stage in writing. You wouldn't expect your first draft of an essay or a business proposal to be perfect. The same is true for prompts.

THE ITERATIVE MINDSET

Here's a truth most people don't like hearing at first: **your first prompt won't be perfect.** And that's okay.

Think of AI like a new colleague who's incredibly talented but doesn't yet know how you like things done. If you gave that colleague a vague assignment and expected flawless results on the first try, you'd probably end up disappointed. But if you view each draft as a conversation — a chance to clarify, guide, and adjust — the results quickly get better. That's what the *iterative mindset* is all about.

SHIFT YOUR EXPECTATIONS

Instead of aiming for "the perfect prompt," expect a first draft. Expect to say, *"This is close, but not quite what I had in mind."* That shift in expectation makes all the difference. Suddenly, "bad" answers aren't failures — they're stepping stones. Each output teaches you how to ask better questions.

STAY CURIOUS, NOT FRUSTRATED

When the AI misses the mark, it's tempting to throw up your hands and think, *"This thing doesn't work."* But iteration asks you to pause and get curious: *Why didn't it work? Did I leave something out? Was I too vague? Did I give conflicting instructions?*

Curiosity keeps you from giving up too soon. Instead, it nudges you to test, tweak, and try again — and often the very next attempt is a huge improvement.

THINK CONVERSATIONALLY

People often treat prompts like magic spells: you write one line, hit enter, and hope for miracles. But AI works best when you treat it like a dialogue. Imagine sitting across the table from a smart collaborator. You wouldn't just blurt out one instruction and walk away. You'd talk through it: "That's good, but can you make it shorter?" or "I like the second point, expand on that." Iteration turns prompting into a back-and-forth, not a one-off command.

FOCUS ON PROGRESS, NOT PERFECTION

The iterative mindset is also about letting go of perfectionism. You don't need a flawless answer — you need one that's useful enough to move forward. Sometimes "good enough" on the third try is better than chasing "perfect" on the twentieth. In fact, many professionals use iteration to quickly generate a solid draft, then polish it themselves. The magic isn't in perfection; it's in progress.

Here's a simple mantra to keep in mind when practicing iteration:

Draft → Adjust → Improve.

MINI-SCENARIOS: THE ITERATIVE MINDSET IN ACTION

To make this concrete, let's look at how iteration plays out for three very different people: a writer, a teacher, and a business leader.

1. The Writer

Samantha, a freelance blogger, asks AI:

- **Prompt 1:** "Write a blog post about productivity."
- **Output:** A generic list of tips like "make a to-do list" and "avoid distractions."

Not very exciting. Instead of giving up, she iterates:

- **Prompt 2:** "Rewrite the blog post with a fun, conversational tone, using humor and relatable examples."
- **Output:** Much better — quirky anecdotes about procrastination and coffee habits.

She refines once more:

- **Prompt 3:** "Keep the humor, but also include two scientific studies on productivity techniques."

Now Samantha has a post that's engaging *and* credible. She didn't need a perfect first try — she needed the willingness to refine.

2. The Teacher

Mr. Alvarez, a 7th-grade history teacher, wants a lesson plan.

- **Prompt 1:** "Write a lesson plan on the American Revolution."
- **Output:** A textbook-style overview with no student engagement.

So he iterates:

- **Prompt 2:** "Rewrite the lesson plan for 7th graders. Include an interactive activity and three discussion questions."
- **Output:** A plan with group debates and critical thinking prompts.

He tweaks again:

- **Prompt 3:** "Simplify the vocabulary so students with lower reading levels can still participate."

Now he has a differentiated lesson plan that meets his students where they are — achieved through iteration, not a one-shot request.

3. The Business Leader

Jordan, a startup CEO, needs a pitch deck.

- **Prompt 1:** "Create slides for an investor pitch."
- **Output:** A bland outline with generic headings.

He refines:

- **Prompt 2:** "Focus the pitch on a fitness app for busy professionals. Highlight market size, competitive advantage, and revenue model."
- **Output:** A much sharper pitch deck.

Still, it feels too formal. One last iteration:

- **Prompt 3:** "Rewrite the pitch to sound bold and visionary, with energetic language and a sense of urgency."

Now Jordan has slides that not only inform but inspire — the kind that get investors leaning in.

REFINEMENT TECHNIQUES

Iteration works best when paired with simple, repeatable strategies:

1. **Re-asking with Adjustments** – Modify small details until the AI locks in.
2. **Progressive Prompting** – Start broad, then zoom in.
3. **Error Correction Prompts** – Point out flaws and request improvements.
4. **Comparison and Contrast** – Ask for multiple versions, then refine the strongest.
5. **Layering Constraints** – Add limits gradually to avoid overload.

Each of these strategies builds momentum and keeps the process focused.

THE FEEDBACK LOOP IN ACTION

Think of iteration as a three-part loop:

1. **Prompt → Output:** You write a prompt, AI generates output.
2. **Evaluation:** You assess what's useful, what's missing, what's off-target.
3. **Refinement:** You adjust the prompt, clarify instructions, or add constraints.

Then the loop repeats. Each cycle gets you closer to your ideal outcome.

WHEN TO RESTART VS. REFINE

Iteration works best when outputs are close but not perfect. Sometimes, though, it's better to restart.

Restart when:

- The output is completely irrelevant (AI misunderstood entirely).
- The prompt is overloaded with contradictions.
- You've iterated so much that the instructions are messy and unclear.

Refine when:

- The output is useful but missing details.
- Tone, style, or scope is slightly off.
- The structure is good but needs polishing.

ADVANCED ITERATION STRATEGIES

- **Meta-Prompting:** Ask the AI to critique its own response.
- **Self-Refinement:** Have the AI generate multiple drafts and improve them.
- **Guided Feedback Loops:** Feed your edits back into the AI for continuity.

These methods let you treat AI as a co-editor, not just a one-time generator.

Common Pitfalls in Iteration

- Endless tweaking that wastes time.
- Over-editing until outputs get worse.
- Discarding useful sections instead of building on them.
- Forgetting to save refined prompts for reuse.

Iteration should simplify, not complicate, your workflow.

CONCLUSION: REFINEMENT UNLOCKS MASTERY

Mastering prompt engineering isn't about writing perfect prompts on the first try. It's about embracing iteration as a process of continual improvement. Each loop of feedback makes your collaboration with AI smarter, sharper, and more efficient.

So next time you use an AI tool, remember:

- The first answer is just a draft.
- Every refinement is progress.
- The feedback loop is your most powerful tool.

In the next chapter, we'll take iteration even further by exploring advanced prompting techniques — including system vs. user prompts, few-shot learning, and persona-based prompting.

CHAPTER 6
ADVANCED PROMPTING TECHNIQUES

UP TO THIS POINT, you've built a solid foundation in prompt engineering — mastering clarity, specificity, context, structure, and iteration. These are the essentials, the equivalent of learning to walk before you run.

Now it's time to take things further. Once you've got the basics down, you can move into advanced techniques that unlock an entirely new level of control and creativity. These are the methods that power users, researchers, and professionals rely on to push AI to its limits.

In this chapter, we'll explore advanced prompting strategies such as system vs. user prompts, few-shot learning, persona-based prompting, chain-of-thought reasoning, self-refinement, and instruction chaining. Along the way, you'll see practical examples, case studies, and mini-exercises that bring each technique to life.

SYSTEM VS. USER PROMPTS

Every interaction with AI involves two layers of instruction:

- **User prompts**: What *you* type in directly.
- **System prompts**: The "background rules" that frame the AI's role before your request.

Most of the time, system prompts are invisible (e.g., "You are a helpful assistant"). But when you gain control over them, they can dramatically shift the output.

Example:

- User-only: "Explain gravity." → Dry, technical definition.
- With system prompt: "You are a patient science teacher who explains concepts to 8-year-olds." User prompt: "Explain gravity." → "Gravity is like an invisible glue that pulls everything down. It's why when you jump, you land back on the ground."

📌 **Guiding Question:**

- How does changing the system prompt shift the AI's style or tone?
- In what situations might you want a formal vs. casual system role?
- What risks exist if you leave the system prompt too vague?

Takeaway: Setting a strong system prompt establishes tone, audience, and perspective before the user prompt even begins.

Mini-Exercise: Write the same user prompt ("Explain how photosynthesis works") with three different system prompts (teacher, comedian, lawyer). Notice how the style shifts dramatically.

FEW-SHOT AND ZERO-SHOT LEARNING

AI models excel at pattern recognition. By giving them examples, you help them "see" the style, tone, or structure you want.

- Zero-shot prompting: You give no examples — just the task.
    - Example: "Translate this sentence into French.
- Few-shot prompting: You give one or more examples to teach the model the pattern you want.
    - Example:

- Input: "Good morning → Bonjour"
- Input: "Thank you → Merci"
- Now translate: "How are you?"

Few-shot prompting leverages the model's ability to learn patterns from context. Even just one or two examples can dramatically improve consistency, so Few-shot prompting can be invaluable when consistency matters — for instance, when you're producing dozens of product descriptions or grading rubrics in the same format.

Use cases:

- Consistent formatting (e.g., "Answer in the style shown in these examples").
- Training the AI to adopt your brand's tone or writing style.
- Demonstrating specialized jargon or technical vocabulary.

📌 **Guiding Question:**

- When would you use zero-shot prompting instead of few-shot prompting?
    - How do examples improve consistency in AI outputs?
    - What types of examples would help the AI match your preferred style or format?

Mini-Exercise: Try writing a zero-shot summary and then a few-shot summary with a clear example. Compare the tone and consistency.

PERSONA-BASED PROMPTING

Assigning the AI a persona transforms how it responds. Instead of generic answers, you get targeted insights that match a role or worldview.

Examples:

- Professional: "Act as a contract lawyer. Review this paragraph and flag potential risks."

- Educational: "Act as a math tutor for middle school students. Explain fractions with simple examples."
- Creative: "Pretend you are Shakespeare writing Instagram captions about coffee."

By layering details into the persona (experience, communication style, audience), you give the AI a "voice" that shapes everything it generates.

📌 **Guiding Question:**

- What professional or creative personas would be most useful for your work?
- How much detail should you add to make the persona effective?
- How can persona-based prompting improve creativity or accuracy compared to generic prompts?

Mini-Exercise: Ask the AI to explain the same topic (e.g., climate change) three times — once as a professor, once as a child, and once as a comedian.

MULTI-STEP REASONING (CHAIN-OF-THOUGHT)

Complex problems require more than quick answers. Chain-of-thought prompting encourages the AI to *show its work.*

Example:

- Prompt: "Step by step, explain how to calculate the area of a triangle with base 10 cm and height 5 cm."
- Output:
    1. Recall formula: Area = ½ × base × height.
    2. Plug in values: ½ × 10 × 5.
    3. Calculate: 25 cm².

The benefit isn't just accuracy — it's transparency. You can see how the AI arrived at its conclusion, making errors easier to catch.

📌 **Guiding Question:**

- Why is it important to see the AI's reasoning process, not just its answers?
- How does chain-of-thought prompting increase trust in outputs?
- When might you avoid chain-of-thought reasoning (e.g., when you just want a quick summary)?

Mini-Exercise: Give the AI a word problem and specifically ask for "step-by-step reasoning." Compare this to a one-shot answer. Which inspires more confidence?

SELF-REFINEMENT AND META-PROMPTING

What if the AI could edit itself? That's exactly what meta-prompting does.

- **Meta-prompting:** "Critique your last answer. What's missing or unclear? Rewrite it with improvements."
- **Self-refinement:** "Write three different introductions. Now select the strongest and polish it."

This technique transforms the AI into both creator and editor. It reduces your workload while producing higher-quality results.

Before-and-After Example:

- First attempt: "Meditation is good for you."
- Meta-prompt: "Revise to include specific health benefits and a motivating tone."
- Refined: "Meditation reduces stress, improves focus, and supports emotional well-being. Just ten minutes a day can reset your mind and body."

📌 **Guiding Question:**

- How can asking the AI to critique itself save you time?
- In what situations would self-refinement be more effective than manual editing?
- What's the benefit of generating multiple drafts before choosing the strongest one?

Instruction Chaining

Big requests can overwhelm AI. Instruction chaining breaks them into logical stages.

Example: Blog Article Development

1. Generate 10 title ideas about healthy eating.
2. Pick the top 3 and explain why they're engaging.
3. Turn the best one into a full outline.
4. Write the first draft of the article.

Each stage builds on the last, giving you control and reducing the chance of confusion.

📌 **Guiding Question:**

- Why might a single overloaded prompt fail where chained instructions succeed?
- How can you break a large project into smaller steps for the AI?
- What advantages does instruction chaining offer for complex workflows?

Mini-Exercise: Take a complex task (e.g., "plan a conference") and split it into at least 3 chained prompts. Compare the final output to asking for everything at once.

CASE STUDY: BUILDING A BUSINESS PLAN WITH ADVANCED TECHNIQUES

Imagine a founder creating a business plan for a new fitness app:

1. **Persona-based:** "You are a venture capitalist. Brainstorm 10 features investors would find appealing."
2. **Few-shot:** Provide one example feature written in pitch-friendly language. Ask for the rest in that style.
3. **Chain-of-thought:** Have the AI reason step by step which features are most marketable.
4. **Meta-prompting:** Ask it to critique its own list and refine it to the best five.
5. **Instruction chaining:** Move from features → revenue model → go-to-market strategy.

The result isn't just a brainstorm, but a structured, investor-ready plan — built through layering advanced techniques.

📌 **Guiding Question:**

- Which advanced techniques did the founder combine most effectively?
- How could you adapt this step-by-step approach to your own projects?
- What risks might arise from skipping one of these steps (e.g., skipping refinement)?

Common Pitfalls

- **Over-engineering:** Stacking too many techniques at once creates confusion.
- **Forgetting the basics:** Clarity and context still matter more than complexity.
- **Losing sight of goals:** Don't let clever techniques distract from your actual need.

CONCLUSION: THE POWER OF ADVANCED PROMPTING

Advanced prompting techniques take you beyond basic competence into mastery. They allow you to:

- Guide tone and style (system prompts, personas).
- Ensure consistency (few-shot).
- Increase accuracy and transparency (chain-of-thought).
- Save time with built-in editing (meta-prompting).
- Tame complexity (instruction chaining).

The beauty is that these techniques don't replace the fundamentals — they **build on them**. By mixing and matching thoughtfully, you'll discover that AI isn't just a tool you use, but a partner you can shape, direct, and refine.

📌 **Guiding Question:**

- Which advanced technique feels most immediately useful to you?
- How might combining two or three techniques improve your workflow?
- What challenges do you anticipate when applying these methods in your field?

In the next chapter, we'll take these advanced skills into the real world and explore how prompt engineering works across different domains — from education and business to healthcare and creative industries.

CHAPTER 7
PROMPT ENGINEERING ACROSS DOMAINS

PROMPT ENGINEERING ISN'T JUST a technical skill — it's a universal practice that adapts to nearly every industry. The foundations stay the same (clarity, context, specificity, structure, and iteration), but how you apply them shifts dramatically depending on the domain.

In this chapter, we'll explore five areas where prompt engineering is already transforming work: business, education, healthcare, coding & data science, and the creative industries. Along the way, you'll see not only examples and challenges, but also guiding questions to help you reflect on how these lessons apply to your own world.

BUSINESS APPLICATIONS

In business, prompt engineering is about efficiency, clarity, and impact. Organizations use AI for drafting reports, analyzing trends, writing marketing copy, and even shaping strategy.

Example Prompt:

"You are a management consultant. Write a 700-word report for a CEO explaining three cost-saving strategies for a mid-sized retail company. Include examples and a short action plan."

With this level of clarity and structure, the AI provides a focused, professional report — something a leader could skim in minutes and take straight to a meeting.

But business prompts also go micro. Marketing teams might say:

"Create five Instagram captions promoting a new eco-friendly clothing line. Keep each under 50 words, use a playful tone, and end with a hashtag."

Both macro (reports) and micro (social copy) demonstrate the same principle: define the audience, context, and constraints.

Challenges:

- Market analysis outputs can reflect data biases.
- Over-reliance on AI for decision-making may sideline human judgment.
- Ensuring that all outputs remain aligned with brand voice requires iteration.

📌 **Guiding Question (Business):**

- How can you make sure AI outputs sound "on-brand" for your organization?
- What risks might arise if prompts for business strategy lack specificity?
- Which of your daily business tasks could be streamlined with a stronger prompt?

EDUCATION APPLICATIONS

Educators and students are using AI to create lesson plans, simplify concepts, and personalize learning. The key is shaping prompts to match age, audience, and learning goals.

Example Prompt:

"Act as a high school history teacher. Create a 45-minute lesson plan on the causes of World War I. Include objectives, key vocabulary, an interactive activity, and five discussion questions."

This gives teachers a ready-made, structured plan — a time-saver that still leaves room for customization.

Students can benefit, too. For example:

"Explain photosynthesis to a 6th grader using simple language and a cooking metaphor."

The output transforms complexity into accessibility.

Challenges:

- AI can generate inaccuracies if prompts don't specify sources.
- Students may overuse AI as a shortcut rather than a learning aid.
- Tone must be carefully set so lessons are age-appropriate.

📌 **Guiding Question (Education):**

- How might prompts differ for teaching high schoolers versus elementary students?
- How can you balance AI-generated teaching material with your own creativity?
- What safeguards could prevent AI misuse in education?

HEALTHCARE APPLICATIONS

Healthcare is a high-stakes domain where clarity and accuracy are critical. Prompt engineering here is often about simplifying technical information, creating accessible communication, or organizing research.

Example Prompt:

"Draft a patient-friendly explanation of high blood pressure test results at a 7th-grade reading level, under 300 words."

This makes complicated medical data understandable and less intimidating.

For research and administration:

"Summarize three peer-reviewed studies on mindfulness-based therapy for chronic pain, published after 2020."

Here, the AI acts as a research assistant, saving time while surfacing recent findings.

Challenges:

- Ethical concerns: prompts must avoid unintentionally producing misleading advice.
- Privacy: never share sensitive patient data with public AI systems.
- Accuracy: every AI output must be double-checked by professionals.

📌 **Guiding Question (Healthcare):**

- When is it appropriate to use AI in healthcare communication — and when is it not?
- How can prompts ensure that patient-facing language stays clear and empathetic?
- What role should human review always play in high-stakes outputs?

CODING AND DATA SCIENCE APPLICATIONS

For developers and analysts, AI is like a junior programmer that can explain, debug, and suggest improvements.

Example Prompt:

"Explain this Python error message in plain terms and suggest three possible fixes: [paste error here]."

This turns cryptic errors into manageable action steps.

Or consider data analysis:

"Write a SQL query to retrieve the top 10 customers by purchase volume from this database schema: [insert schema]."

The output gives analysts a shortcut, but still requires review for efficiency and accuracy.

Challenges:

- Code generated may be inefficient or insecure if not reviewed.
- Beginners risk becoming dependent instead of learning fundamentals.
- Without precision, prompts can produce bloated, irrelevant code.

📌 **Guiding Question (Coding):**

- How specific does a coding prompt need to be to avoid useless results?
- In what ways can AI support learning versus replacing it?
- How can you ensure AI-generated code meets security standards?

CREATIVE INDUSTRY APPLICATIONS

Writers, marketers, and artists use AI to spark ideas, brainstorm slogans, or generate drafts. Prompt engineering here means balancing constraint with freedom.

Example Prompt:

"Write a 500-word short story about a detective in 1920s Paris who discovers that he is the criminal he's been chasing. Use a noir style."

This combines constraints (setting, style, twist) with enough room for creativity.

For marketing:

"Generate 10 catchy slogans for a nonprofit focused on ocean conservation, each under 10 words."

The outputs provide multiple options, each a springboard for refinement.

Challenges:

- Too-generic prompts can yield cliché ideas.
- Creative professionals must ensure originality and avoid plagiarism risks.
- Clients' needs still require human refinement beyond AI drafts.

📌 **Guiding Question (Creative):**

- How can prompts balance freedom for creativity with necessary constraints?
- What risks arise when creative prompts are too vague?
- How might you use AI as a brainstorming partner rather than a final creator?

THE CROSS-DOMAIN ADVANTAGE

What's powerful about prompt engineering is that skills learned in one field often transfer to another. A teacher's clarity improves business memos. A developer's precision sharpens storytelling prompts. A marketer's focus on audience context enhances patient communication.

The more you practice across domains, the more versatile you become.

📌 **Guiding Question (Cross-Domain):**

- Which domain's prompting style feels closest to your current work?
- How could you borrow strategies from another field to strengthen your own prompts?
- What risks or blind spots might you miss if you only practice prompting in one domain?

CASE STUDY: A DAY IN THE LIFE OF CROSS-DOMAIN PROMPTING

Imagine Lisa, a consultant who uses AI across multiple domains in one day:

> **1. Business:** Market analysis for a client.
> Prompt: "Act as a market analyst. Summarize renewable energy investment trends since 2020 with three supporting statistics."
> **2. Education:** Helping her nephew with math.
> Prompt: "You are a tutor. Explain fractions to a 5th grader using pizza as an analogy."
> **3. Healthcare:** Drafting a nonprofit health guide.
> Prompt: "Write a 400-word plain-language guide explaining why regular blood pressure checks matter."
> **4. Creative:** Brainstorming slogans.
> Prompt: "Write 10 two-word slogans for a mental health awareness campaign."

By applying the same principles across domains, Lisa multiplies her impact — not by reinventing the wheel each time, but by refining the universal art of prompt engineering.

CONCLUSION: DOMAIN SHAPES DIRECTION

Prompt engineering is universal, but domain shapes direction. Business needs persuasion, education needs adaptability, healthcare needs accuracy, coding needs precision, and creative work needs surprise.

By practicing across domains, you become a more versatile communicator, problem-solver, and creator. The techniques you've learned so far are not confined to one area — they're portable, adaptable, and endlessly useful.

CHAPTER 8
COMMON MISTAKES AND HOW TO AVOID THEM

PROMPT ENGINEERING CAN FEEL deceptively simple. After all, you're just typing words into a box, right? But anyone who has tried knows the truth: the way you phrase your instructions can mean the difference between a brilliant output and a confusing mess.

Even experienced users fall into traps. Some write prompts that are too vague, others overload with details, and many forget to iterate. These aren't failures — they're part of the learning curve. The good news is that once you recognize these mistakes, you can correct them quickly and unlock far better results. In this chapter, we'll walk through the 10 most common mistakes in prompt engineering, with examples, fixes, and guiding questions to help you reflect.

MISTAKE 1: BEING TOO VAGUE

Vague prompts are the fastest path to generic outputs.

Example:

"Write about climate change."

The result is almost guaranteed to be broad, shallow, and uninspiring.

Better Prompt:

"Write a persuasive 500-word op-ed arguing that local governments should prioritize renewable energy projects over fossil fuel subsidies. Include two recent U.S. statistics (after 2020) and a motivational closing statement."

This prompt gives the AI a *purpose (persuade)*, *format (op-ed)*, *scope (renewable vs. fossil fuels)*, and *constraints (length and data)*.

📌 **Guiding Questions:**

- Have you noticed AI giving you generic outputs? Could vagueness in your prompt be the cause?
- Which details — audience, purpose, format — do you most often forget to include?

MISTAKE 2: OVERLOADING THE PROMPT

If vagueness is one extreme, **overloading** is the other.

Example:

"Act as a sustainability consultant. Write a 750-word report for executives on reducing carbon emissions, with three case studies, five bulleted action steps, a glossary of terms, an appendix formatted for Excel, and a motivational closing quote."

The AI will try, but the response will likely be cluttered, incomplete, or inconsistent.

Fix: Break the task into stages (instruction chaining). For example:

1. Generate case studies.
2. Draft the bulleted steps.
3. Add a glossary.
4. Assemble into a polished report.

📌 **Guiding Questions:**

- Have you ever stuffed too many requests into one prompt? What happened?
- Which parts of your complex tasks could you separate into smaller steps?

MISTAKE 3: CONTRADICTORY INSTRUCTIONS

Contradictions confuse the AI and dilute results.

Example:

"Write in a formal academic tone, but also keep it casual and funny."

That's like asking a chef to make food both spicy and bland at the same time.

Fix: Prioritize or request multiple versions.

"Write two versions: one in a formal academic tone, another in a light-hearted style."

📌 **Guiding Questions:**

- Have you ever asked for two styles at once? What did the AI do with that?
- How could you make your priorities clearer when you have mixed needs?

MISTAKE 4: FORGETTING CONTEXT

AI doesn't know your situation unless you tell it.

Example:

"Summarize inflation."

Without context, the AI may give you a textbook-style explanation.

Better Prompt:

"Explain inflation to a high school economics class in under 300 words, using two real-world examples that affect teenagers (like rising food or gas prices)."

📌 **Guiding Questions:**

- Do you assume AI already knows your audience?
- Which contexts (audience, purpose, situation) do you most often forget to include?

MISTAKE 5: NEGLECTING CONSTRAINTS

Without boundaries, outputs often sprawl.

Example:

"Write an article on social media."

The AI might produce a long, unfocused essay.

Better Prompt:

"Write a 200-word article summarizing the top three social media platforms among U.S. teens in 2023. Keep the tone professional and concise."

Constraints (word count, focus, tone) make the difference.

📌 **Guiding Questions:**

- Have you ever received outputs that were far too long or irrelevant?
- What types of constraints (length, style, focus) would have prevented that?

MISTAKE 6: SKIPPING ITERATION

Too many users expect perfection in one shot.

Example:

Prompt: "Write ad copy for a fitness app."

Output: Generic and uninspired.

User: Frustrated, gives up.

Fix: Iterate. For example:

- Round 1: "Make it more motivational, aimed at busy professionals."
- Round 2: "Add a strong call-to-action under 20 words."

By Round 3, you'll have something polished.

📌 **Guiding Questions:**

- Do you expect AI to get it right the first time?
- How could treating outputs as drafts change your results?

MISTAKE 7: IGNORING ROLES

Without a role, outputs default to generic.

Example:

"Write safety guidelines for a construction site."

Better Prompt:

"You are a workplace safety officer. Write bulleted guidelines for new construction workers on protective gear and hazard awareness."

Roles focus the AI's perspective and voice.

📌 **Guiding Questions:**

- What roles could make your prompts more effective?
- How might a doctor, lawyer, or teacher persona change the response?

MISTAKE 8: IGNORING LIMITATIONS

Some users assume AI is an all-knowing oracle.

Example:

"Give me the exact success rate of a new cancer drug."

Without context, the AI might invent numbers.

Fix: Ask for sources or add constraints.

"Summarize clinical trial findings on [drug name] since 2021. Include references from peer-reviewed journals."

📌 **Guiding Questions:**

- Have you ever trusted an AI answer without fact-checking?
- Which types of prompts demand careful validation before use?

MISTAKE 9: FORGETTING THE AUDIENCE

AI may produce correct information in the wrong style or complexity.

Example:

"Explain blockchain."

Too abstract for a child, too simple for an expert.

Better Prompt:

"Explain blockchain to a 12-year-old using an analogy about trading cards."

📌 **Guiding Questions:**

- Who is your real audience when you use AI?
- How can you ensure prompts reflect their needs and level of knowledge?

MISTAKE 10: NOT SAVING GOOD PROMPTS

Many people rediscover the same prompts again and again.

Fix: Build a **prompt library.** Save, categorize, and reuse the ones that work best.

📌 **Guiding Questions:**

- Do you have a system for saving your best prompts?
- How could a personal "prompt playbook" save you hours in the long run?

CASE STUDY: LEARNING THROUGH MISTAKES

Brian, a project manager, struggled with AI early on.

- **Round 1:** He asked, "Help me with a report." → Output: vague.
- **Round 2:** He overloaded: "Write a 1,000-word report with case studies, bulleted lists, data tables, and references." → Output: incoherent.
- **Round 3:** He broke it into steps: generate key points → expand into sections → polish into a report. → Output: clear, useful, professional.

Brian's lesson? Mistakes aren't failures. They're part of the process.

📌 **Guiding Questions:**

- Which of Brian's mistakes have you made yourself?
- How might breaking tasks into stages help you avoid overload or vagueness?

CONCLUSION: FROM PITFALLS TO POWER

Every mistake in prompt engineering points to a skill you can strengthen. Vague prompts teach you to be clearer. Overloaded prompts teach you to simplify. Ignoring audience teaches you to adapt. Trusting AI too much and finding errors reminds you to fact-check.

Prompt engineering isn't about avoiding mistakes entirely — it's about **learning through them.**

📌 **Guiding Questions (Final Reflection):**

- Which mistake do you recognize most often in your own practice?
- What's one small change you could make this week to avoid it?
- How might creating a "common mistakes checklist" keep you on track?

CHAPTER 9
ETHICS AND RESPONSIBILITY IN PROMPT DESIGN

AS AI BECOMES part of everyday workflows, the way we design prompts isn't just a technical exercise — it's also an ethical one. A single poorly phrased prompt can unintentionally spread misinformation, amplify bias, or misuse sensitive information. On the other hand, carefully crafted prompts can foster fairness, clarity, and accountability. Ethical prompt design means asking not only *"Will this work?"* but also *"Should this be used in this way?"* and *"Who might be impacted by the output?"*

In this chapter, we'll explore five key ethical considerations in prompt engineering: bias and fairness, accuracy and misinformation, privacy and data security, harmful or manipulative use, and transparency and accountability.

BIAS AND FAIRNESS

AI doesn't create ideas from scratch. AI systems reflect the data they are trained on. If the training data skews toward particular demographics, regions, or cultural patterns, the outputs can unintentionally reinforce those biases. This isn't always malicious, but it is consequential. A biased prompt or output can shape perceptions, decisions, and opportunities in ways that disadvantage entire groups, unintentionally reinforcing negative stereotypes.

Example of Hidden Bias:

Prompt: *"List the top inventors in history."*

Likely output: A long list of white, male inventors from Western contexts, while ignoring female, non-Western, or Indigenous innovators.

Better Prompt:

"List 10 inventors in history from diverse cultural, gender, and geographic backgrounds. Include at least three women and at least three non-Western innovators."

Here, the prompt itself acts as a fairness check.

Methods for Reviewing Outputs for Bias

1. **Representation Scan:** After receiving an output, ask: *Whose voices are missing? Which groups are overrepresented?*
2. **Counter-Prompting:** Re-run the same prompt but explicitly request diversity (e.g., "Include African, Asian, and Indigenous perspectives"). Compare results.
3. **Perspective Shift:** Ask AI to reframe from another angle: *"How would this answer look if written from the perspective of a woman in STEM?"*
4. **Bias Testing Prompts:** Use diagnostic prompts such as:
   - "Does this response reflect stereotypes? If so, how could they be avoided?"
   - "What groups or perspectives are missing from this answer?"

📌 **Guiding Questions (Bias & Fairness):**

- How often do you review AI outputs for missing perspectives?
- Could you add fairness checks into your prompts before running them?

- How might you use counter-prompting to surface hidden bias?

ACCURACY AND MISINFORMATION

One of the most dangerous issues in AI-generated text are outputs that sound authoritative but are factually wrong, typically referred to as AI "hallucinations." This happens because AI generates output based on patterns, not on an internal fact-checking mechanism. Without careful prompting and validation, misinformation can spread quickly.

Example of Misinformation Risk:

Prompt: *"What were the results of the most recent Nobel Prize in Medicine?"*

If the model's training cut-off predates the award, it may confidently invent a winner.

Better Prompt:

"Summarize the most recent Nobel Prize in Medicine. If your knowledge cutoff prevents access to the 2023 award, state that clearly."

This approach prompts the AI to admit uncertainty instead of fabricating.

Methods for Reviewing Outputs for Accuracy

1. Source Requesting: Always ask: *"Provide references with URLs from peer-reviewed or reputable outlets."*
2. Cross-Verification: Check whether key claims can be verified with at least two trusted sources (e.g., academic journals, government sites).
3. Uncertainty Prompts: Include in your prompt: *"If unsure, say so instead of guessing."*
4. Fact-Check Loops: After receiving output, run a follow-up: *"List three possible errors or gaps in the above response."*
5. Domain Expert Review: If the content is high-stakes (e.g., medical, financial, legal), always have a human expert validate before use.

📌 **Guiding Questions** (Accuracy & Misinformation):

- When do you need to demand sources or citations from AI? and when do you skip this step?
- How can you design prompts that encourage honesty about uncertainty?
- Which verification methods (cross-checking, counter-prompting, expert review) would fit best in your work?

PRIVACY AND DATA SECURITY

One of the most common ethical mistakes is pasting sensitive or confidential data into public AI systems. Even if the system says it won't store the data, you may still risk breaches.

Example:

Prompt: "Summarize this patient's medical record and suggest a treatment plan."

This is a major privacy violation if used in an unapproved AI system.

Better Prompt:

Keep sensitive data out of prompts. Instead, use anonymized placeholders:

"Summarize this anonymized medical case: [generalized information]. Highlight three treatment options based on recent guidelines."

📌 **Guiding Questions (Privacy):**

- Have you ever pasted confidential information into AI without thinking?
- How could you reframe sensitive data into anonymized or synthetic examples?
- What policies should organizations create for safe AI use?

HARMFUL OR MANIPULATIVE USE

Prompts can be designed to manipulate, harass, or even spread harmful content. While AI systems have guardrails, prompt engineers have a responsibility not to try to "jailbreak" those safeguards.

Example:

Prompt: "Write a realistic fake news article blaming a political leader for a disaster."

Even if blocked, the intention itself crosses ethical lines.

Better Prompt:

Instead of using AI for manipulation, use it for critical thinking:

"Analyze how fake news articles are written. Identify common rhetorical strategies and suggest ways to teach media literacy."

📌 **Guiding Questions (Harmful Use):**

- Have you seen prompts that intentionally aim to mislead or manipulate?
- How can you distinguish between analyzing harmful content (for research/education) and generating it?
- Where should you draw the line when experimenting with edgy prompts?

TRANSPARENCY AND ACCOUNTABILITY

When using AI, it's tempting to present outputs as entirely your own work. But in many fields — academia, journalism, healthcare — failing to acknowledge AI contributions can be misleading.

Example:

Prompt: "Write a research paper introduction on climate change."

If you copy-paste without attribution, readers may think it's fully original.

Better Prompt:

Use AI as a collaborator, not a ghostwriter:

"Draft an outline for a research paper introduction on climate change. I'll refine and add my own citations."

Transparency doesn't always mean citing AI as an author — but it does mean being honest about when and how you used it.

📌 **Guiding Questions (Transparency):**

- In your field, when should you disclose AI assistance?
- How might hiding AI involvement damage credibility?
- What guidelines could help distinguish between "AI as a tool" vs. "AI as an author"?

CASE STUDY: AN ETHICAL CROSSROADS

Imagine Priya, a marketing manager under pressure to deliver.

- **Scenario 1:** She pastes customer email lists directly into AI to generate personalized copy. Quick, but a privacy breach.
- **Scenario 2:** She asks AI to write persuasive copy for her campaign but notices the examples reinforce gender stereotypes.
- **Scenario 3:** She uses AI to write the entire press release, then publishes it without edits or disclosure.

Each decision carries risks. The ethically responsible path is to:

1. Protect customer privacy (use anonymized or generic personas).
2. Add fairness constraints ("avoid stereotypes, use inclusive language").
3. Edit and fact-check outputs, disclosing AI's role in her workflow when appropriate.

📌 **Guiding Questions (Case Study):**

- Which of Priya's missteps feels most common in your own work?
- How could she have designed prompts differently to reduce risk?
- What's the balance between speed, creativity, and ethical responsibility in your context?

PRACTICAL CHECKLIST FOR ETHICAL PROMPTING

1. **Check for Bias** → Am I reinforcing stereotypes or excluding groups?
2. **Check for Accuracy** → Did I ask for sources or fact-check outputs?
3. **Check for Privacy** → Did I paste sensitive or confidential data?
4. **Check for Harm** → Could this output be misused to manipulate or harm?
5. **Check for Transparency** → Should I disclose that AI was part of this process?

CONCLUSION: THE RESPONSIBLE PROMPT ENGINEER

Ethics in prompt design isn't about making your work harder. It's about making your work more trustworthy, fair, and sustainable. Every time you write a prompt, you're not just shaping an output — you're shaping the way people experience information. The best prompt engineers don't just ask, *"What can I make the AI do?"* They also ask, *"What's the impact of this output, and who could it affect?"*

📌 **Guiding Questions (Final Reflection):**

- Which ethical risk — bias, accuracy, privacy, harm, or transparency — feels most relevant to your work?
- What's one ethical "rule of thumb" you could adopt starting today?

- How might you train your team or peers to recognize these same responsibilities?

CHAPTER 10
THE FUTURE OF PROMPT ENGINEERING - FROM ART TO AUTOMATION

PROMPT ENGINEERING TODAY feels like a blend of art and science — a skill that combines creativity, clarity, and structured thinking. But where is it headed? Will prompt engineers always need to carefully craft words, or will tools and models eventually automate much of this work?

In this chapter, we'll explore the emerging trends shaping the future of prompt engineering: evolving model capabilities, automation and "promptless" AI, standardization and best practices, integration with other technologies, and the shifting role of humans in the loop.

EVOLVING MODEL CAPABILITIES

Language models are rapidly improving. Early AI systems struggled with vague prompts, but newer models are better at inferring context and handling ambiguity. This raises a question: will prompt engineering become less necessary as models get "smarter"? The answer is nuanced. While advanced models reduce the burden of micromanaging every detail, clear and responsible prompting will remain essential for guiding outputs toward fairness, accuracy, and usefulness.

📌 **Guiding Questions (Model Evolution):**

- How do newer models reduce the need for hyper-specific prompts?
- In what areas (creative writing, technical problem-solving, decision support) do you think human-crafted prompts will remain indispensable?
- How could improved models change the balance between creativity and control?

AUTOMATION AND PROMPTLESS AI

We're already seeing the rise of automated prompt generation. Tools like AI "prompt optimizers" rewrite vague user inputs into more detailed instructions before passing them to the model. Other systems, like AI assistants built into apps, anticipate user needs without explicit prompting.

Imagine asking your AI calendar: *"Can you help me get organized for next week?"* Instead of needing a structured prompt, the system infers your intent, pulls events, drafts reminders, and even prioritizes tasks. This is the shift from prompt engineering to orchestration — where prompts are layered, optimized, and automated behind the scenes.

📌 **Guiding Questions (Automation):**

- What risks might arise if users no longer see or understand the prompts happening behind the scenes?
- How could automated prompting improve accessibility for non-experts?
- What guardrails should exist to prevent misuse of invisible prompt automation?

STANDARDIZATION AND BEST PRACTICES

As industries adopt AI, we'll likely see **prompt engineering standards** emerge — guidelines for clarity, fairness, and transparency, much like today's coding or UX standards.

For example, hospitals may establish a rule: *"All prompts must anonymize patient data and require citation from peer-reviewed sources."* Businesses may define brand-specific tone frameworks for prompts.

Universities, governments, and companies may even certify "prompt design practices," ensuring ethical and effective use.

📌 **Guiding Questions (Standards):**

- What kinds of prompt standards would benefit your industry most?
- How might checklists (for accuracy, privacy, bias) be integrated into daily workflows?
- Could prompt design become a professional certification, like project management or coding?

INTEGRATION WITH OTHER TECHNOLOGIES

Prompt engineering won't exist in isolation. Future workflows will likely integrate AI prompting into other tools:

- **Voice Interfaces:** Prompts spoken naturally instead of typed.
- **Multimodal Systems:** Prompts mixing text, images, and data tables.
- **Agent Systems:** AI handling multi-step tasks across apps, guided by a chain of prompts you never see.

For instance, an architect might upload a floor plan, then say: *"Generate three sustainable design variations."* The AI integrates images, data, and natural language seamlessly.

📌 **Guiding Questions (Integration):**

- How might multimodal prompts (mixing images, text, and data) expand your work?
- What opportunities and risks do voice- or agent-based prompting create?
- How could integrated AI change how teams collaborate?

THE SHIFTING ROLE OF HUMANS

If AI systems can anticipate and optimize prompts, does that mean prompt engineers will become obsolete? Not exactly. Instead, the human role will shift from writer to curator, auditor, and ethicist.

Humans will:

- **Set boundaries:** Ensuring outputs align with ethical and organizational values.
- **Audit outputs:** Checking for bias, misinformation, and relevance.
- **Direct creativity:** Asking questions that machines can't anticipate.
- **Translate intent:** Bridging between human goals and AI execution.

Prompt engineering will move closer to strategic communication and design thinking — less about writing "magic words" and more about shaping systems responsibly.

📌 **Guiding Questions (Human Role):**

- How do you see your role evolving as AI gets better at interpreting intent?
- What human qualities (judgment, empathy, creativity) will remain irreplaceable?
- How might prompt engineering merge with other roles (like project management or ethics review)?

FROM ART TO AUTOMATION: A BALANCED FUTURE

So, what's the future? Prompt engineering won't vanish — it will evolve. The art of carefully shaping words will coexist with automated systems that reduce friction. Standards will formalize best practices. Human oversight will remain critical for ethics and accountability.

The ultimate shift is from **art to orchestration**:

- Today → Crafting prompts by hand.
- Tomorrow → Guiding ecosystems of AI that handle prompts behind the scenes.

This means the best prompt engineers of the future won't just be wordsmiths — they'll be system designers, strategists, and ethical leaders.

CASE STUDY: PROMPT ENGINEERING IN 2030

Picture Maya, a "Prompt Architect" in 2030.

- At work, she oversees an AI platform that drafts policy briefs for government agencies. Instead of writing prompts herself, she reviews the automated prompt pipelines for fairness, accuracy, and transparency.
- At home, her smart home assistant anticipates her needs: "Looks like you're hosting dinner Friday — do you want me to draft a shopping list and playlist?" No explicit prompt needed.
- In her community, she teaches students how to critically evaluate AI outputs, empowering them to ask better questions and challenge biases.

Maya isn't typing prompts line by line — she's shaping systems of prompts, trust, and responsibility.

📌 **Guiding Questions (Case Study):**

- Which part of Maya's future role feels most relevant to you?
- How could you prepare now for an AI future where prompts are automated?
- What skills (beyond writing prompts) will you need to thrive in this environment?

THE FUTURE IS HYBRID

Prompt engineering is at a crossroads. It began as an art of crafting words and is moving toward automation, orchestration, and oversight.

Your role as a prompt engineer may change, but your value will not disappear. Instead, it will shift from "How do I phrase this?" to "How do I shape this system responsibly, creatively, and ethically?"

The future belongs to those who see prompt engineering not just as a skill, but as part of a broader ecosystem of communication, design, and accountability.

📌 **Guiding Questions (Final Reflection):**

- Do you see prompt engineering in your future as more art, more automation, or both?
- How can you start preparing now for AI tools that may prompt themselves?
- What role do you want to play in shaping ethical and effective AI systems?

CHAPTER 11
PROMPT ENGINEERING SUCCESS STORIES

THROUGHOUT THIS BOOK, we've explored the principles, frameworks, and ethics of prompt engineering. But theory only goes so far. To truly understand the transformative potential of prompt design, we need to look at how real people and organizations have used it to achieve measurable success.

In this chapter, we'll highlight four domains — Business, Finance, Healthcare, and Education — and examine stories that show how the art and science of prompting is not just abstract, but practical and powerful. Each story underscores the same lesson: well-crafted prompts drive better outcomes.

BUSINESS: GORGIAS + PROMPTLAYER (E-COMMERCE / CUSTOMER SUPPORT AUTOMATION)

One of the clearest real-world examples comes from Gorgias, a customer support platform used by many SaaS and e-commerce businesses. Gorgias uses PromptLayer to help its clients scale AI-driven support while managing prompt engineering workflows.

Here's how it works in practice:

- The support team defines standardized prompt templates (role-based, constraint-focused) that align with their brand voice and policies.
- Every new support ticket that can be automated is fed through these prompt templates (with dynamic variables like customer name, issue category).
- PromptLayer tracks prompt versions, measures performance (e.g. resolution rate, deflection vs escalation), and lets engineers refine prompts iteratively.

In practice, Gorgias reports that automation expanded across support with less oversight, freeing human agents for critical, high-touch cases. Because prompt versioning and evaluation were built in, they avoided regressions (i.e. a prompt that used to work doesn't degrade over time).

VERIZON'S GENAI FOR CUSTOMER LOYALTY

Verizon is using generative AI internally to predict why customers call, match them to the right agents, and proactively reduce churn. Though the public narrative doesn't detail the prompt designs, the project likely involves **constraint-driven prompts** and **role-based prompts** to interpret conversation context, identify intent, and route calls appropriately.

The reported results:

- Predicting reasons for ~80% of calls
- Reducing store visit time by ~7 minutes per customer
- Improving retention by addressing issues faster

This shows how prompt engineering underlies large-scale AI use in telecommunications.

How these business examples tie to prompt engineering principles:

- Templating & standardization make automation reliable.
- Monitoring and iteration help maintain performance over time.
- Role + constraints guide the AI to act in brand-aligned ways.

📌 **Guiding Questions (Business):**

- Where in your organization could reusable prompt templates save time?
- How could prompts enforce brand voice across different teams or regions?
- What risks arise if prompts aren't carefully aligned with company values?

FINANCE: INTELLIGENT REASONING & TRANSPARENCY

While fewer public case studies talk explicitly about "prompt engineering," we can look at how companies are adopting AI in fintech and investment with reasoning-guided systems and infer how prompt design plays a role.

Klarna's AI Chatbot Handling 2/3rds of Support Queries

The Swedish fintech firm Klarna now uses an AI chatbot (developed in collaboration with OpenAI) to manage about two-thirds of its customer service inquiries. The bot handles multiple languages across many jurisdictions, and the company claims that in its first month it handled 2.3 million conversations.

This suggests strong prompt engineering behind the scenes, such as:

- Few-shot prompts to guide tone and format across regions
- Constraint prompts for length, compliance, and risk
- Iteration cycles to tune responses for various financial regulations

The business impact is profound: lowered support cost, faster resolutions, and scale across territories with minimal incremental staffing.

Comcast's "Ask Me Anything" Agent Assist System

Comcast engineers have created an "AMA" (Ask Me Anything) tool embedded into agent workflows. It allows agents to query a language model mid-conversation. The system uses prompt engineering to generate context-aware suggestions or answers.

In experiments, agents who used AMA had ~10% less time spent per conversation that required a search, leading to millions of dollars in annual saving. Because AMA is tightly integrated, its prompt templates need to adapt in real-time (i.e. initial user message, agent context, conversation history).

These finance / service-context examples illustrate how prompt engineering underpins efficiency and transparency in high-stakes domains.

FINANCE: ENHANCING RISK ANALYSIS WITH CHAIN-OF-THOUGHT PROMPTS

A different mid-size investment firm struggled with synthesizing massive volumes of market data. Analysts were wasting hours pulling trends, summarizing reports, and drafting briefs. The firm piloted a large language model to help — but early outputs were shallow and inconsistent. The breakthrough came when the analysts applied Chain-of-Thought prompting. Instead of simply asking for summaries, they instructed the AI to "explain reasoning step by step, highlight assumptions, and flag uncertainties."

Example Prompt (for financial trend analysis):

"Analyze these quarterly earnings reports. Step by step, identify key revenue drivers, note risks, and explain how they connect to broader market trends. Highlight at least two assumptions you are making and where data is incomplete."

The results:

- Greater accuracy. Analysts caught errors by following the AI's reasoning trail.
- Time savings. Draft market briefs were cut from 6 hours to 1.5 hours.
- Improved compliance. Documented reasoning supported regulatory reviews.

The firm didn't eliminate human expertise. Instead, prompt engineering elevated analysts to decision-makers rather than data wranglers.

📌 **Guiding Questions (Finance):**

- Would AI Chatbots improve your customer service performance?
- How could Chain-of-Thought prompts make reasoning more transparent in your field?
- What compliance or audit benefits might step-by-step prompting provide?
- How would you balance AI-generated insights with human judgment?

HEALTHCARE: CLINICAL DOCUMENTATION & AI SCRIBES

Healthcare is one of the domains where prompt design matters deeply because of the stakes involved: privacy, accuracy, and patient safety. Healthcare professionals spend a staggering share of their time on administrative work. A U.S. hospital network tested AI tools to automate clinical note drafting — but the real leap came from refining the prompts.

Heidi Health: AI Medical Scribe & Documentation Automation

Heidi Health, an Australian health tech company, offers an AI scribe platform that transcribes clinical encounters into structured notes, case histories, and medical documents. Their system uses LLMs plus domain-specific prompt design to convert spoken or transcribed conversations into usable clinical documentation.

Key elements of their prompting architecture include:

- Using prompt templates aligned with clinical note structures (SOAP or narrative)
- Constraining output so that only medically relevant details are included
- Including safety and privacy guardrails to avoid PHI leakage

Because of this, clinicians report that substantial portions of their notes are drafted automatically, saving time and improving documentation consistency.

Sporo AI Scribe vs GPT-4o Mini Study

A comparative study evaluated Sporo AI Scribe and GPT-4o Mini in generating clinical documentation. Researchers used prompt engineering methodologies — instructing models to produce SOAP-style notes and guiding them to include intermediate steps or clarifying assumptions.

Results:

- Sporo's prompt-tuned model outperformed GPT-4o Mini in recall, precision, and overall F1 scores (which measure accuracy of output).
- Clinicians rated Sporo's outputs higher for accuracy, comprehensiveness, and relevance.
- Fewer "hallucinations" (false statements) when prompts included constraints and reasoning instructions.

This reinforces how carefully designed prompts (not just more compute) drive better clinical outputs.

These real examples show that in healthcare, prompt engineering can transform burdensome tasks into trustworthy support tools — when done with structure, governance, and domain alignment.

📌 **Guiding Questions (Healthcare):**

- What existing frameworks (like SOAP) could you embed into prompts for clarity?
- How could prompts be designed to reduce risk of including sensitive data?
- Which administrative burdens in your field might be ripe for prompt-driven support?

EDUCATION: PERSONALIZED LEARNING WITH ADAPTIVE PROMPTS

While publicly documented education-specific prompt engineering examples are fewer, the principles transfer. Imagine a university using prompts like:

- "Act as a tutor in calculus. Review this student's solution, identify two mistakes, explain them step-by-step, and suggest one practice problem."
- "Create three differentiated lesson plans for geometry: for beginners, intermediates, and advanced students, each with challenge questions."

Some AI educational platforms implicitly use versions of these prompts behind their adaptive content engines. And many educators have become early adopters / creative prompt engineers of their own volition. One high school teacher in Canada wanted to provide more individualized feedback to students but struggled with limited time. She began using AI with role-based and constraint-focused prompts to create differentiated learning materials.

Example Prompt (for personalized feedback):

"Act as a supportive teacher. Give feedback on this student essay at a Grade 10 reading level. Identify two strengths and two growth areas.

Keep your response under 200 words, and encourage the student positively."

Example Prompt (for lesson design):

"Design three math practice problems on quadratic equations for a student struggling with factoring. Provide step-by-step solutions and one encouraging tip."

By iterating and refining, the teacher built a library of prompt templates for feedback, lesson differentiation, and practice material generation.

The results:

- Students received timely, personalized feedback.
- Teacher time was preserved, enabling her to focus on live instruction.
- Student engagement improved, with learners reporting they felt "seen" and supported.

The teacher later shared her prompt library with colleagues, multiplying the impact across the school.

📌 **Guiding Questions (Education):**

- How might prompts help you scale personalized support in your work?
- What tone or role constraints could ensure positive, constructive outputs?
- How could you adapt a "prompt library" for your team or classroom?

LESSONS ACROSS DOMAINS

Though these stories come from very different fields, a few patterns emerge:

1. **Frameworks matter.** Whether SOAP in healthcare or Chain-of-Thought in finance, anchoring prompts to established structures boosts trust and clarity.
2. **Constraints shape quality.** Limiting word count, requiring tone, or mandating format turns generic text into actionable output.
3. **Iteration is essential.** Each story involved trial, error, and refinement before prompts became truly effective.
4. **Humans remain central.** Success came not from outsourcing judgment but from empowering people with better tools.

FROM STORIES TO STRATEGY

Prompt engineering is not just theory. Across industries, people are using it to save time, improve accuracy, reduce burnout, and scale personalization. These stories show what's possible when prompts are treated not as throwaway inputs but as designed artifacts — tools crafted with care, creativity, and responsibility.

Your own success story could be next. The challenge is to take the principles you've learned in this book and apply them in your context — experimenting, refining, and documenting what works.

CONCLUSION: BECOMING A RESPONSIBLE PROMPT ENGINEER

We began this book with a simple truth: the words you choose shape the outputs you receive. Over the course of these chapters, we've traveled from the basics of clarity and structure to the complexities of bias, automation, and ethics. What started as an introduction to "how to talk to AI" has grown into something bigger: a framework for responsible, thoughtful, and future-ready communication with intelligent systems. As we close, let's tie together the threads of this journey: what prompt engineering means today, how it will evolve tomorrow, and how you can carry these lessons forward.

Key Lessons to Carry Forward

1. Clarity Is Power.
Every strong prompt starts with being clear about audience, purpose, and format. When in doubt, ask yourself: *What exactly am I asking for, and why?*

2. Context Shapes Relevance.
Without context, outputs drift into generic territory. With context, they become precise, tailored, and actionable.

3. Constraints Are Creative.
Far from limiting, constraints sharpen outputs. Word counts, formats, tones — these are the rails that guide creativity.

4. Iteration Beats Perfectionism.
Expecting the perfect answer on the first try sets you up for frustration. Treating outputs as drafts transforms AI into a collaborator.

5. Ethics Aren't Optional.
Bias, misinformation, privacy, manipulation, transparency — these aren't afterthoughts. They are the heart of responsible prompting.

6. The Future Is Hybrid.
Prompt engineering will evolve from manual craft to automated orchestration. Your role will shift from wordsmith to auditor, designer, and ethical guide.

YOUR ROLE AS A PROMPT ENGINEER

Whether you're a teacher, business leader, coder, or creative, prompt engineering is more than a technical trick — it's a form of digital literacy. In the same way that learning to read, write, or code opened doors in past generations, the ability to craft, critique, and oversee AI interactions will define opportunities in the decades ahead.

The role you play will depend on your context:

- **In business,** you may use prompts to streamline communication, analyze markets, or shape customer experiences.
- **In education,** prompts may help you design lessons, simplify complexity, or spark curiosity.
- **In healthcare,** prompts must balance clarity with accuracy, always guided by professional oversight.
- **In creative industries,** prompts become catalysts for new ideas and imaginative breakthroughs.

But across all these domains, the common thread is **responsibility**. Each prompt is an act of authorship — not just of text, but of influence.

PRACTICAL NEXT STEPS

1. Build Your Prompt Library.
Save and categorize the prompts that work well for you. Over time, you'll create a personal playbook of strategies.
2. Experiment Across Domains.
Don't limit yourself to one use case. Try prompts in business, education, coding, and creativity. Transfer skills across contexts.
3. Audit for Bias and Accuracy.
Regularly ask: *Whose voices are missing? What facts need checking?* Build review loops into your process.
4. Practice Iteration Daily.
Treat AI as a dialogue, not a vending machine. Experiment, refine, and adjust. Make feedback a habit.

5. Stay Informed.
The field of AI is evolving rapidly. New models, methods, and ethical guidelines emerge constantly. Commit to ongoing learning.

LOOKING AHEAD

Prompt engineering is still young. In a few years, some aspects may be automated, standardized, or integrated into everyday tools. But the essence — the ability to shape human-AI collaboration through intentional communication — will endure.

The future belongs not to those who use AI passively, but to those who engage it critically, creatively, and responsibly.

📌 **Guiding Questions (Final Reflection):**

- Which lessons from this book resonate most with your daily work?
- How will you apply the principles of clarity, context, constraints, iteration, and ethics in your own prompts?
- What kind of role do you want to play in shaping the future of human-AI collaboration?

A CLOSING THOUGHT

Prompt engineering began as an art of words, but it is becoming an art of systems — an art of designing not just outputs, but trust. Every time you write a prompt, you are shaping a micro-world: a conversation, a story, a piece of knowledge, or a decision. To practice prompt engineering responsibly is to recognize that these micro-worlds add up. They ripple outward — into classrooms, boardrooms, clinics, and communities.

And so, the final call is simple: become not just a prompt engineer, but a responsible one. Carry forward the clarity, creativity, and care that make this practice not just useful, but meaningful. Because the future of prompt engineering isn't just about better AI. It's about building a better partnership between humans and the systems we create.

STUDY & DISCUSSION GUIDE

FOR UNDERSTANDING PROMPT *Engineering*

This guide compiles the **Guiding Questions** from every chapter into one place. Use it for self-reflection, journaling, classroom study, or team discussions. Each set of questions is designed to help you apply the lessons of the book to your own context.

How to Use This Guide

- **Self-Study:** Pick 1–2 questions per chapter and journal your answers.
- **Team Discussions:** Use these as prompts for group dialogue about how AI affects your work.
- **Workshops:** Assign sections of the book, then explore the corresponding questions together.
- **Classrooms:** Adapt as discussion starters or short essay prompts.

The goal isn't to answer every question perfectly — it's to keep the **conversation about prompt engineering active, reflective, and practical.**

Chapter 1: The Power of Prompts – Why Words Shape AI Outputs

- How does the quality of a prompt influence the quality of the AI's output?
- What differences did you notice between the "bad" and "better" prompt examples?
- In your own work, where could clearer instructions make collaboration (with AI or people) more effective?
- How can prompts act as both instructions and creative sparks?

Chapter 2: Core Principles of Effective Prompting

- When writing prompts, how can you check for clarity before submitting?
- How does specificity reduce irrelevant or unfocused AI outputs?
- Why is context just as important for AI as it is for human communication?
- In what situations might too much structure limit creativity?
- How could you practice combining clarity, specificity, context, and structure into one strong prompt?

Chapter 3: Prompt Structures and Frameworks That Work

- Which framework feels most natural for you to use (Role–Task–Context–Output, Step-by-Step, Chain-of-Thought, Few-shot, Persona)?
- How can breaking tasks into steps improve both AI accuracy and your own efficiency?
- What risks come from overloading a single prompt with too many requirements?
- How might you adapt scaffolding prompts to your own projects (e.g., writing, teaching, business planning)?
- Which real-world tasks could benefit most from instruction frameworks?

Chapter 4: Context and Constraints – Setting Boundaries for Better Results

- Why does audience context matter when shaping AI outputs?
- How can constraints improve the usefulness of AI responses?
- What happens when a prompt has too many constraints?
- Which tasks in your daily work could benefit most from adding background context?
- How can you balance freedom and boundaries when crafting prompts?

Chapter 5: Iteration and Refinement – Mastering the Feedback Loop

- Why is it helpful to treat AI outputs as drafts instead of final answers?
- How does curiosity improve the iteration process?
- In your own projects, how could treating prompts conversationally lead to better results?
- Which of the mini-scenarios (writer, teacher, business leader) felt most relatable, and why?
- What strategies (error correction, progressive prompting, layering constraints) could you test this week to practice refinement?
- How can you tell when it's better to restart a prompt vs. refine it?

Chapter 6: Advanced Prompting Techniques

- How does changing the system prompt shift the AI's style or tone?
- When would you use zero-shot prompting instead of few-shot prompting?
- What professional or creative personas would be most useful for your work?
- Why is it important to see the AI's reasoning process, not just its answers?
- How can asking the AI to critique itself save you time?

- Why might a single overloaded prompt fail where chained instructions succeed?
- Which advanced technique feels most immediately useful to you?
- How might combining two or three techniques improve your workflow?

Chapter 7: Prompt Engineering Across Domains

- How do prompting strategies shift between business, education, healthcare, coding, and creative fields?
- Which domain examples feel closest to your own work?
- How can practicing prompts across different domains make you stronger overall?
- What risks are unique to high-stakes domains like healthcare or law?
- How might you borrow prompting techniques from another field to improve your own?

Chapter 8: Common Mistakes and How to Avoid Them

- When was the last time you got a vague AI output? Could the issue have been your prompt?
- How do you know when a prompt is "overloaded"?
- Have you ever asked for two styles at once? What did the output look like?
- What contexts do you often leave out of your prompts?
- Have you ever received outputs that were far too long or too short?
- Do you expect AI to get it "right" on the first try?
- What roles could make your prompts more effective?
- Have you ever assumed an AI's answer was correct without checking?
- Who is your real audience when you use AI?
- Do you have a system for saving your best prompts?
- Which of Brian's mistakes (case study) felt familiar to you?

Chapter 9: Ethics and Responsibility in Prompt Design

- How might your prompts unintentionally reinforce stereotypes?
- What constraints could you add to ensure diverse or inclusive outputs?
- When is it appropriate to explicitly mention fairness in your prompt?
- When do you need to demand sources or citations from AI?
- How could vague prompts increase the risk of misinformation?
- Have you ever pasted confidential information into AI without thinking?
- How can you distinguish between analyzing harmful content and generating it?
- In your field, when should you disclose AI assistance?
- Which of Priya's missteps (case study) feels most common in your own work?
- Which ethical risk (bias, accuracy, privacy, harm, transparency) feels most relevant to your work?

Chapter 10: The Future of Prompt Engineering – From Art to Automation

- How do newer models reduce the need for hyper-specific prompts?
- In what areas do you think human-crafted prompts will remain indispensable?
- What risks might arise if users no longer see prompts happening behind the scenes?
- What kinds of prompt standards would benefit your industry most?
- How might multimodal prompts (mixing images, text, data) expand your work?
- How do you see your role evolving as AI gets better at interpreting intent?
- Which part of Maya's future role (case study) feels most relevant to you?

- Do you see prompt engineering in your future as more art, more automation, or both?

Chapter 11: Prompt Engineering Success Stories

- Where in your organization could reusable prompt templates save time?
- How could prompts enforce brand voice across different teams or regions?
- What risks arise if prompts aren't carefully aligned with company values?
- Would AI Chatbots improve your customer service performance?
- How could Chain-of-Thought prompts make reasoning more transparent in your field?
- What compliance or audit benefits might step-by-step prompting provide?
- How would you balance AI-generated insights with human judgment?
- What existing frameworks (like SOAP) could you embed into prompts for clarity?
- How could prompts be designed to reduce risk of including sensitive data?
- Which administrative burdens in your field might be ripe for prompt-driven support?
- How might prompts help you scale personalized support in your work?
- What tone or role constraints could ensure positive, constructive outputs?
- How could you adapt a "prompt library" for your team or classroom?

PROMPT LIBRARY

HOW TO USE **Your Prompt Library**

1. **Duplicate and Save:** Keep a folder or binder where you store completed templates.
2. **Refine Over Time:** Add notes, refinements, and iterations.
3. **Tag & Organize:** Label by domain (business, education, creative, coding, healthcare).
4. **Reuse & Adapt:** Good prompts are reusable — adapt them to new contexts.

Clarity / Basic Prompt

Task/Goal:	
Audience:	
Tone/Style:	
Format Needed (list, essay, summary, etc.):	
Example Prompt:	
Notes/Refinements:	

Task/Goal:	
Audience:	
Tone/Style:	
Format Needed (list, essay, summary, etc.):	
Example Prompt:	
Notes/Refinements:	

Task/Goal:	
Audience:	
Tone/Style:	
Format Needed (list, essay, summary, etc.):	
Example Prompt:	
Notes/Refinements:	

Task/Goal:	
Audience:	
Tone/Style:	
Format Needed (list, essay, summary, etc.):	
Example Prompt:	
Notes/Refinements:	

Role-Based Prompts

Role-Based Prompt

Assigned Role (Teacher, Analyst, Journalist, etc):	
Task:	
Audience:	
Constraints (tone, word limit, focus areas):	
Example Prompt:	
Output Quality Check:	☐ Matches role voice ☐ Clear and relevant ☐ Adapted to audience

Assigned Role (Teacher, Analyst, Journalist, etc):	
Task:	
Audience:	
Constraints (tone, word limit, focus areas):	
Example Prompt:	
Output Quality Check:	☐ Matches role voice ☐ Clear and relevant ☐ Adapted to audience

Assigned Role (Teacher, Analyst, Journalist, etc):	
Task:	
Audience:	
Constraints (tone, word limit, focus areas):	
Example Prompt:	
Output Quality Check:	☐ Matches role voice ☐ Clear and relevant ☐ Adapted to audience

Step-by-Step Prompt

Overall Goal:	
Key Steps:	1. 2. 3.
Constraints/ Requirements:	
Example Prompt:	
Follow-Up Prompt for Next Step:	

Overall Goal:	
Key Steps:	1. 2. 3.
Constraints/ Requirements:	
Example Prompt:	
Follow-Up Prompt for Next Step:	

Overall Goal:	
Key Steps:	1. 2. 3.
Constraints/ Requirements:	
Example Prompt:	
Follow-Up Prompt for Next Step:	

Chain-of-Thought Prompt

Problem/Question:	
Instructions for Reasoning:	☐ Explain step-by-step reasoning ☐ Show intermediate steps ☐ Highlight assumptions
Example Prompt:	
Output Review:	☐ Logical flow? ☐ Easy to follow? ☐ Matches intended depth?

Problem/Question:	
Instructions for Reasoning:	☐ Explain step-by-step reasoning ☐ Show intermediate steps ☐ Highlight assumptions
Example Prompt:	
Output Review:	☐ Logical flow? ☐ Easy to follow? ☐ Matches intended depth?

Problem/Question:	
Instructions for Reasoning:	☐ Explain step-by-step reasoning ☐ Show intermediate steps ☐ Highlight assumptions
Example Prompt:	
Output Review:	☐ Logical flow? ☐ Easy to follow? ☐ Matches intended depth?

Problem/Question:	
Instructions for Reasoning:	☐ Explain step-by-step reasoning ☐ Show intermediate steps ☐ Highlight assumptions
Example Prompt:	
Output Review:	☐ Logical flow? ☐ Easy to follow? ☐ Matches intended depth?

Few-Shot Prompt

Task:	
Audience:	
Desired format:	
Examples Provided:	1. 2. 3.
Example Prompt with Few-Shot Input	
Evaluation:	☐ Output matches examples ☐ Tone/style consistent ☐ Accurate and usable

Task:	
Audience:	
Desired format:	
Examples Provided:	1. 2. 3.
Example Prompt with Few-Shot Input	
Evaluation:	☐ Output matches examples ☐ Tone/style consistent ☐ Accurate and usable

Task:	
Audience:	
Desired format:	
Examples Provided:	1. 2. 3.
Example Prompt with Few-Shot Input	
Evaluation:	☐ Output matches examples ☐ Tone/style consistent ☐ Accurate and usable

Persona-Based Prompt

Persona Description (who, traits, background):	
Task for Persona:	
Tone/Style (casual, formal, humorous, etc.):	
Constraints:	
Review Checklist:	☐ Persona voice consistent ☐ Relevant to audience ☐ Creative but within bounds

Persona Description (who, traits, background):	
Task for Persona:	
Tone/Style (casual, formal, humorous, etc.):	
Constraints:	
Review Checklist:	☐ Persona voice consistent ☐ Relevant to audience ☐ Creative but within bounds

Persona Description (who, traits, background):	
Task for Persona:	
Tone/Style (casual, formal, humorous, etc.):	
Constraints:	
Review Checklist:	☐ Persona voice consistent ☐ Relevant to audience ☐ Creative but within bounds

Persona Description (who, traits, background):	
Task for Persona:	
Tone/Style (casual, formal, humorous, etc.):	
Constraints:	
Review Checklist:	☐ Persona voice consistent ☐ Relevant to audience ☐ Creative but within bounds

Constraint-Focused Prompt

Task:	
Audience:	
Constraints:	WORD COUNT_____ Style/Tone: _____ Format: _____
Example Prompt:	
Output Check:	☐ Within word count ☐ Matches required tone/style ☐ Delivers specified format

Task:	
Audience:	
Constraints:	WORD COUNT_____ Style/Tone: _____ Format: _____
Example Prompt:	
Output Check:	☐ Within word count ☐ Matches required tone/style ☐ Delivers specified format

Task:	
Audience:	
Constraints:	WORD COUNT_____ Style/Tone: _____ Format: _____
Example Prompt:	
Output Check:	☐ Within word count ☐ Matches required tone/style ☐ Delivers specified format

Task:	
Audience:	
Constraints:	WORD COUNT_____ Style/Tone: _____ Format: _____
Example Prompt:	
Output Check:	☐ Within word count ☐ Matches required tone/style ☐ Delivers specified format

Iterative / Refinement Prompt

Initial Prompt:	
Output Received (summary):	
Feedback / Refinement Request:	
Second Iteration Prompt:	
Notes on Improvement:	
Final Version (after iterations):	

Initial Prompt:	
Output Received (summary):	
Feedback / Refinement Request:	
Second Iteration Prompt:	
Notes on Improvement:	
Final Version (after iterations):	

Initial Prompt:	
Output Received (summary):	
Feedback / Refinement Request:	
Second Iteration Prompt:	
Notes on Improvement:	
Final Version (after iterations):	

Initial Prompt:	
Output Received (summary):	
Feedback / Refinement Request:	
Second Iteration Prompt:	
Notes on Improvement:	
Final Version (after iterations):	

Multimodal / Integration Prompt

Task:	
Inputs (image, table, dataset, etc.):	
Output Needed (caption, summary, analysis, design idea):	
Constraints/Instructions:	
Example Prompt:	
Review Checklist:	☐ Integrated input correctly ☐ Matches requested output ☐ Accurate + creative as needed

Task:	
Inputs (image, table, dataset, etc.):	
Output Needed (caption, summary, analysis, design idea):	
Constraints/Instructions:	
Example Prompt:	
Review Checklist:	☐ Integrated input correctly ☐ Matches requested output ☐ Accurate + creative as needed

Task:	
Inputs (image, table, dataset, etc.):	
Output Needed (caption, summary, analysis, design idea):	
Constraints/Instructions:	
Example Prompt:	
Review Checklist:	☐ Integrated input correctly ☐ Matches requested output ☐ Accurate + creative as needed

Grounded Context Prompt

Retrieved Context To Be Used:	
Prompt Variables: (audience/ format/ tone/ style)	
Question(s) Related to Context:	
Notes/Refinements:	

Retrieved Context To Be Used:	
Prompt Variables: (audience/ format/ tone/ style)	
Question(s) Related to Context:	
Notes/Refinements:	

Retrieved Context To Be Used:	
Prompt Variables: (audience/ format/ tone/ style)	
Question(s) Related to Context:	
Notes/Refinements:	

Retrieved Context To Be Used:	
Prompt Variables: (audience/ format/ tone/ style)	
Question(s) Related to Context:	
Notes/Refinements:	

Retrieved Context To Be Used:	
Prompt Variables: (audience/ format/ tone/ style)	
Question(s) Related to Context:	
Notes/Refinements:	

REFERENCES

- Bender, E. M., Gebru, T., McMillan-Major, A., & Shmitchell, S. (2021). On the dangers of stochastic parrots: Can language models be too big? *Proceedings of the 2021 ACM Conference on Fairness, Accountability, and Transparency*, 610–623. https://doi.org/10.1145/3442188.3445922
- Bommasani, R., Hudson, D. A., Adeli, E., Altman, R., Arora, S., von Arx, S., ... & Liang, P. (2021). *On the opportunities and risks of foundation models. arXiv preprint.* https://arxiv.org/abs/2108.07258
- Brown, T., Mann, B., Ryder, N., Subbiah, M., Kaplan, J., Dhariwal, P., ... & Amodei, D. (2020). *Language models are few-shot learners. Advances in Neural Information Processing Systems*, 33, 1877–1901. https://arxiv.org/abs/2005.14165
- Clark, J. (2023). *The art of the prompt: How clarity, context, and specificity shape AI outputs. Journal of Applied AI Research,* 5(2), 45–61.
- Dale, R. (2021). GPT-3: What's it good for? *Natural Language Engineering,* 27(1), 113–118. https://doi.org/10.1017/S1351324920000601
- Floridi, L., & Cowls, J. (2019). A unified framework of five principles for AI in society. *Harvard Data Science Review,* 1(1). https://doi.org/10.1162/99608f92.8cd550d1
- Gao, L., Tow, J., & Zhang, A. (2023). Pal: Program-aided language models. *arXiv preprint arXiv:2211.10435.* https://arxiv.org/abs/2211.10435
- Jobin, A., Ienca, M., & Vayena, E. (2019). The global landscape of AI ethics guidelines. *Nature Machine Intelligence,* 1(9), 389–399. https://doi.org/10.1038/s42256-019-0088-2
- Liu, P., Yuan, W., Fu, J., Jiang, Z., Hayashi, H., & Neubig, G. (2023). Pre-train, prompt, and predict: A systematic survey of prompting methods in natural language processing. *ACM Computing Surveys,* 55(9), 1–35. https://doi.org/10.1145/3560815
- Mishra, S., Khashabi, D., Baral, C., & Hajishirzi, H. (2022). Cross-task generalization via natural language crowdsourcing instructions. *Proceedings of the 60th Annual Meeting of the Association for Computational Linguistics (Volume 1: Long Papers),* 3470–3487. https://doi.org/10.18653/v1/2022.acl-long.243
- OpenAI. (2022). *ChatGPT: Optimizing language models for dialogue.* https://openai.com/blog/chatgpt
- OpenAI. (2023). *Our approach to AI safety.* https://openai.com/safety
- OpenAI. (2023). *Our approach to AI safety.* https://openai.com/safety
- Perez, E., Kiela, D., & Cho, K. (2021). True few-shot learning with language models. *Advances in Neural Information Processing Systems,* 34, 11054–11070.
- Reynolds, L., & McDonell, K. (2021). Prompt programming for large language models: Beyond the few-shot paradigm. *arXiv preprint.* https://arxiv.org/abs/2102.07350

- Schick, T., & Schütze, H. (2021). Exploiting cloze-questions for few-shot text classification and natural language inference. *Proceedings of the 16th Conference of the European Chapter of the Association for Computational Linguistics: Main Volume*, 255–269. https://doi.org/10.18653/v1/2021.eacl-main.20
- Wei, J., Wang, X., Schuurmans, D., Bosma, M., Ichter, B., Xia, F., ... & Zhou, D. (2022). Chain-of-thought prompting elicits reasoning in large language models. *arXiv preprint arXiv:2201.11903*. https://arxiv.org/abs/2201.11903
- White, J., Fu, Q., Hays, S., Sandborn, P., Olea, C., Gilbert, S., & Schmidt, D. C. (2023). A prompt pattern catalog to enhance prompt engineering with ChatGPT. *arXiv preprint*. https://arxiv.org/abs/2302.11382

www.ingramcontent.com/pod-product-compliance
Lightning Source LLC
Chambersburg PA
CBHW072022060426
42449CB00034B/1656